Building Production and Project Management

R A Burgess and G White

Building Production and Project Management

R A Burgess and G White

The Construction Press

LANCASTER   **LONDON**   **NEW YORK**

The Construction Press Ltd.,
Lancaster, England.

A subsidiary company of Longman Group Ltd., London.
Associated companies, branches and representatives
throughout the world.

Published in the United States of America by
Longman Inc. New York.

First published 1979.

ISBN 0 904406 93 8

Printed in Great Britain at The Pitman Press, Bath.

Contents

Preface

The idea of this book grew from the authors' experience as examiners in the Final Examination of the Institute of Builders and from many years of management eduction in the construction industry. Whilst its coverage corresponds closely to most academic and professional syllabi in the field of Building Production and Project Management, other related areas are included as any treatment in too narrow a field must inevitably result in only a partial understanding of the context in which a manager may be required to function.

This book is not intended to break new ground in its approach to construction management, nor does it set out to be deliberately provocative or to base its thinking on lengthy academic theory. Rather, it is hoped that it will provide the embryo construction manager with a soundly based foundation upon which to build his expertise in terms which he can recognise as applicable to the situations which he is likely to meet in his everyday experience.

However, it should not be assumed that such an objective dictates an elementary treatment of the subject, as there is much of a more advanced nature that will interest the experienced construction manager in helping to update and extend his knowledge and in many cases challenge his established ideas. Care has been taken to avoid any extensive use of quantitative theories, in fact the maxim has been simple figures for complex ideas. Nor was it felt desirable to include more than a mention of computor-based processes, since these are in so rapid a state of change at this present time, and it was further considered that any discussion in depth would not help towards a clearer understanding of basic construction management principles.

The book is published for the international market and it is hoped that students as far apart as Canada, Africa and Malaysia will find the language of the book acceptable. It is for this reason that references to British Standards and other documents peculiar to the U.K. situation have been kept to the minimum necessary for a clear understanding of the message of the work.

Finally the authors would like to express their thanks to Mrs D E Sweet for helpful comments and much of the typing of the manuscript and to Bob Nye for his valuable assistance with the diagrams.

Roger A. Burgess
Gordon White

1. Management and the Design Process

The function of design

The design process is the essential preliminary activity for all building and may take place within or outside the contractor's organisation. It is normally a function which is divided between several offices each responsible for its own specialised aspect of the process. Thus it will be seen that it has characteristics which create a complexity in its management structure demanding good methods of co-ordination and communication.

One of the hazards in the management of the design process is the difficulty of assessing, with some degree of certainty, the likely duration of the stages of the work, or indeed the effort required of the design team. This in turn means that the cost is also liable to considerable variation and is not necessarily easily predicted.

A further variable factor of some importance is the level of completeness required of the design information at each stage of the design so that other participants may be able to carry out their own activities without the danger of inconsistencies of interpretation or incompatibility of design decisions, and that construction may start.

Whether the design is carried out by a separate team of design consultants (mainly architects and engineers) or by a sub-section of the constructor's own staff, the design processes which are to be co-ordinated follow a very similar pattern which is largely dictated by the present organisation of the industry where the interface between design and production has long been recognised as extremely flimsy despite much that has been written and spoken in attempts to fuse both functions more tightly together. This general process is that which is set out most clearly in the RIBA Plan of Work.

There has been much thought given to the best methods of approaching the solution of design problems and there is a strong school of thought which supports the foundation of a quantitative basis for design technology, possibly closely connected to the interactive use of computers. Whilst the thinking which underlies this work is both exciting and strikes at the restrictions to creative problem solving which are to be found in the traditional design methods, it has only a comparatively small, though increasing, number of followers among practitioners in the architectural profession, the main advances taking place in the engineering industry or in academic establishments. Eventually many changes will come about, largely due to the pressure of more readily accessible computer facilities, but at present there are indeed few deviations to be found from the RIBA Plan of Work which has already been mentioned. Design management control procedures are usually based upon improving and refining this established sequence of events rather than changing fundamentally the method of working.

The manner in which this chapter deals with the subject is to accept in general principle the Plan of Work and to examine the best way to control each stage in terms of time, cost and resource allocation, whilst also considering the use of computers in the design process and their implications for the future in so far as they may influence the management function.

The design process

Many definitions of the design process have been written, varying in their approach according to the background experience from which they are written. For example, the electrical engineer will anticipate greater need for quantification than the architect, who must combine the need for measured data on environmental control with the attributes of personal judgement necessary in the study and interpretation of human needs.

Two definitions seem appropriate here. The first is written by Morris Asimow, an engineer, in his book "Introduction to Design" (1962), which was the forerunner of many subsequent works defining a morphology of design on the basis of systems engineering. He describes design almost entirely in terms of information processes. He writes that the design process consists of:

"the gathering, handling and creative organising of information relevant to the problem situation; it prescribes the derivation of decisions which are optimised, communicated and tested or otherwise evaluated; it has an iterative character, for often, in the doing, new information becomes available or new insights are gained which require the repetition of earlier operations."

It should be noted to what an extent he emphasises the concepts of *information* and its *communication*, as these are the fundamental functions of the designer.

The second source which has been chosen for quotation is the management pundit, Peter Drucker, who in his book "Practice of Management" outlines the design activities as comprising:

1. Defining the problem

2. Analysing the problem

3. Developing alternative solutions

4. Deciding upon the best of these alternatives

5. Communicating the design decisions in such a way that they result in effective action.

Thus he sees the process as one of *problem identification and solution.* Indeed, it may be compared to a process of learning in which one explores a broad problem and thus gradually identifies its detailed facets. Unfortunately the design process has not the defined body of knowledge usually available to guide the learner in his explorations in other fields. In design, he must recall information acquired from his own and others' past experience and gradually refine his decisions as his knowledge builds up. Ultimately he will arrive at the realisation of when conflicts exist between the requirements of different building users, and by his judgement and understanding he will come to a suitable compromise which, as a design solution, offers a balance between the range of requirements to be answered.

From this it will be seen that the three stages of design consist of:

(a) analysis

(b) synthesis

(c) evaluation.

Page has pointed out that, whilst resembling the central phases of decision sequences, these stages do not, in fact, form a single simple sequence which can be followed through from analysis to evaluation. The design process is cyclical, and it is only by reaching one stage of synthesis that the designer realises he has forgotten to analyse something else, which in turn will lead to a modified synthesis. And so on through the cycle!

This is a very important point and one which tends to be overlooked frequently in the formulation of models of the design process and in the management systems which purport to control that process. It is those management procedures with which we are here concerned rather than the relative merits of the schools of design methods which have developed systems intended to introduce a more systematic approach whilst recognising its cyclical nature and the difficulties of quantifying adequate data to solve the infinite range of problems likely to be encountered.

The stages of design   We are therefore accepting the three general stages of design as those already identified, which comprise:

1. *Analysis:* the determination of objectives or goals; the identification of problems and difficulties; exploring relationships between parts of the problem and producing order from random information.

2. *Synthesis:* the procedure of creating solutions for the various parts of the problems, grouping these together into feasible overall solutions whilst generating original ideas.

3. *Evaluation:* the testing of alternative solutions against appropriate selected criteria in order to establish those which meet the requirements most adequately.

We must now consider the management function in the light of the design of buildings, following the Plan of Work as defined in the RIBA Management Handbook (1967), which sets out the key tasks at each stage of the design process. These are illustrated in Table 1.1, from which it can be seen that stages A to F embrace the design functions which in the general sequence of building precede the activities of tendering and appointing a contractor (G and H) and, of course, those of construction (J-M).

It is this sequence of events which forms the background of this chapter, though it is realised that there are many variations of the contractual system which may bring forward some stages (for example, G and H) and subdivide others. However, the tasks of management still remain to be carried out, and it is only the responsibilities for these functions which may alter under different contractual systems. Fundamentally it is a responsibility to foresee and solve those problems which may arise in each stage and which may be summarised as follows:

(a) to ensure that all the technical and other information is available for each stage.

(b) to provide the right blend of professional skills to resolve all the problems which can be anticipated.

(c) to make sure that everyone concerned in the project understands the extent and nature of their responsibilities.

(d) to see that good communication links exist between all those concerned and that they are properly used.

(e) to ensure that all decisions are timely and are made known to those whom they affect.

Table 1.1. Outline plan of work

Stage	Purpose of work and decisions to be reached	Tasks to be done	People directly involved	Usual Terminology
A. Inception	To prepare general outline of requirements and plan future action.	Set up client organisation for briefing. Consider requirements, appoint architect.	All client interests, architect.	Briefing
B. Feasibility	To provide the client with an appraisal and recommendation in order that he may determine the form in which the project is to proceed, ensuring that it is feasible, functionally, technically and financially.	Carry out studies of user requirements, site conditions, planning, design, and cost, etc, as necessary to reach decisions.	Clients' representatives, architects, engineers, and QS according to nature of project.	
C. Outline proposals	To determine general approach to layout, design and construction in order to obtain authoritative approval of the client on the outline proposals and accompanying report.	Develop the brief further. Carry out studies on user requirements, technical problems, planning, design and costs, as necessary to reach decisions.	All client interests, architects, engineers, QS and specialists as required.	Sketch plans
D. Scheme design	To complete the brief and decide on particular proposals, including planning arrangement appearance, constructional method, outline specification, and cost, and to obtain all approvals.	Final development of the brief, full design of the project by engineers, preparation of cost plan and full explanatory report. Submission of proposals for all approvals.	All client interests, architects, engineers, QS and specialists and all statutory and other approving authorities.	

Brief should not be modified after this point.

Stage	Purpose of work and decisions to be reached	Tasks to be done	People directly involved	Usual Terminology
E. Detail design	To obtain final decision on every matter related to design, specification, construction and cost.	Full design of every part and component of the building by collaboration of all concerned. Complete cost checking of designs.	Architects, QS, engineers and specialists, contractor (if appointed).	Working drawings

Any further change in location, size, shape, or cost after this time will result in abortive work.

Stage	Purpose of work and decisions to be reached	Tasks to be done	People directly involved	Usual Terminology
F. Production information	To prepare production information and make final detailed decisions to carry out work.	Preparation of final production information, ie, drawings, schedules and specifications.	Architects, engineers and specialists, contractor (if appointed).	
G. Bills of Quantities	To prepare and complete all information and arrangements for obtaining tender.	Preparation of Bills of Quantities and tender documents.	Architects, QS, contractor (if appointed).	
H. Tender action	Action as recommended in paras 7-14 inclusive of 'Selective Tendering'.*	Action as recommended in paras 7-14 inclusive of 'Selective Tendering'.*	Architects, QS, engineers, contractor, client.	
J. Project planning	Action in accordance with paras 5-10 inclusive of 'Project Management'.*	Action in accordance with paras 5-10 inclusive of 'Project Management'.*	Contractor, sub-contractors.	Site operations
K. Operations on site	Action in accordance with paras 11-14 inclusive of 'Project Management'.*	Action in accordance with paras 11-14 inclusive of 'Project Management'.*	Architects, engineers, contractors, sub-contractors, QS, client.	
L. Completion	Action in accordance with paras 15-18 inclusive of 'Project Management'.*	Action in accordance with paras 15-18 inclusive of 'Project Management'.*	Architects, engineers, contractor, QS, client.	
M. Feed-back	To analyse the management, construction and performance of the project.	Analysis of job records. Inspections of completed building. Studies of building in use.	Architect, engineers, QS, contractor, client.	

Publication of National Joint Consultative Council of Architects, Quantity Surveyors and Builders.

The design team

The design is usually in the hands of a team of professional specialists either employed in the one office, or brought together from separate enterprises as necessary. The client, of course, must also be considered as one of the team, which will normally comprise:

Client with his specialist advisers
Overall Project Manager (ie, Project Co-ordinator)
Architectural Designer
Quantity Surveyor
Civil and Structural Engineers
Building Services Engineers

and it may also include

Contractor
Landscape Architect
Interior Designer
Town Planner.

The function of co-ordination is a vital element in the success of the operation of such a team and this is well stated by Caudill when he writes of such a team as *"an association of people who share common goals to create architecture, who are willing to co-operate and who can communicate with each other"*.

From the foregoing comments it can now be seen that the Design Manager needs to combine successfully two different functions, which may be loosely defined as:

1. *The management function:* to ensure that the project as a whole is well run, and to co-ordinate the process of design.

2. *The design function:* to contribute his own design skills in the solution of the problems and in the making of judgements.

The present structure of the building industry encourages a particular process of design and production to be followed with little interface between the two, and this may be illustrated as shown in Figure 1.1.

Figure 1.1   **Interaction between design and production**

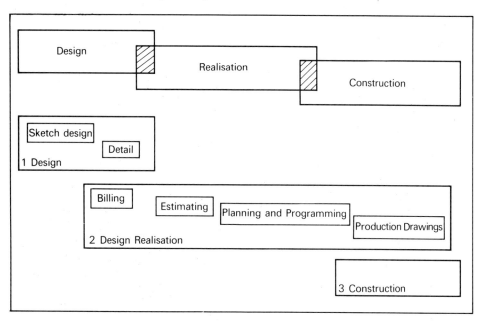

The RIBA Handbook has reduced the design process to its simplest terms as a flow system (Figure 1.2) which represents a main progression, or sequence, with occasional feedback. In reality, the sequence is not a steady flow, as has already been noted, but rather a series of jumps forward interspersed by references back for reconsideration and re-examination. It is this spasmodic nature of the flow of events which makes its control so difficult and its timing so unpredictable. When it is realised that such a sequence is made up of the inter-related activities being carried out by each member of the design team, the importance of the communication system becomes obvious. Further, it must be recalled that each member may be awaiting design decisions before finalising his own, or may be unaware that the information he has been given is now already being reconsidered and may even have been revised.

Figure 1.2   **The design process**

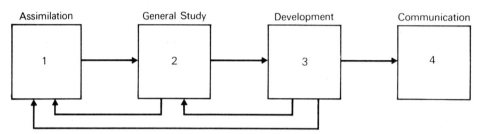

In practice, the mental activity tends to make short flashes from one phase to another according to results achieved and the ideas that are stimulated by the work.

The process of design may be conveniently divided into two phases which correspond to stages in the Plan of Work:

1. Stages A-D, the compilation and completion of the brief and its interpretation in the form of a design solution acceptable to the client's organisation, together with obtaining of all formal approvals.

2. Stages E, F and G, the determination of all final decisions related to design, specification, construction and cost, and the preparation of the production information for the execution of these decisions, in the form of drawings, specification and schedules.

The organisational framework for controlling each phase will differ in so far as the client's contribution in the early stages is the dominating influence which declines in importance in the second stage when any variation in the brief is to be avoided. Therefore the second phase is controlled by the design team's own monitoring system. The role played by the client (or rather the complexity and professionalism of many client systems) depends upon the size and experience of the client; in the case, say, of a hospital, the system will be headed by a person who has overall responsibility for the client's contribution, supported by a number of adminstrative officers, whose joint roles will be to represent the client on all matters from initiation of the project to its completion. These officers will act as the liaison channels for all the inputs from the client side to the design process, by ensuring that the various sections within the client's organisation provide the information and finance the job. The information in the brief will be drawn from the technical, social and administrative requirements of the client which may well be of a highly specialised and often contradictory nature, so that reconciliation between two needs becomes a matter of discussion and tactful analysis of priorities in the resolution of which the design manager's experience can be a vital factor.

On the other hand the client may be a small inexperienced group, or an individual, for whom the definition of his needs is a new path to be trodden, and who will look to his professional design advisers for help and guidance, even if the project to the experienced man may appear quite straightforward and simple. In such cases it is nearly true to say that the client's principal function is to provide the finance at the appropriate times.

The design team will be headed by the design manager, who will co-ordinate and control the work of the various design disciplines involved in each stage of the project, through liaison with a leader appointed within each discipline. In the case of building design, the manager, or leader, of the team is normally an architect who has overall responsibility for the project with a team of designers working under him, or a project architect who will be responsible for the day-to-day decisions and actions of the team of architects and technicians.

Control systems

The control systems to be used for each of the main phases of the work must now be considered. The mechanics by which the first phase is executed consist in the main of a series of meetings and discussions in order to develop the brief, analyse the requirements which the design is to serve and consider technical problems. The more complex the job, the greater is likely to be the variety of topics embraced by such meetings. For example, these could well include such matters as site availability and its potential for development, access, town and country planning requirements, availability of essential services such as water, electricity, telephone and drainage, roads and highways, specialist equipment, the necessary degree of internal design flexibility, and security. Table 1.2 gives some examples.

Table 1.2   **Information for building**

Specific information

The site
Whether it has already been acquired or whether further advice is to be given after inspection by the architect
Whether any restrictions or easements exist, or whether any conditions are attached to the freehold or leasehold
The availability of public services, or whether these will have to be provided privately
Whether the site is complete, or will be extended at some future date, its present area and the area of the future acquisition

Accommodation
Areas and relation of principal rooms or units
Height of rooms
Natural or artificial lighting and ventilation
Internal and external finishings
Population of the building and numbers of either sex to be accommodated
Whether any sections of the building are to be used separately or at different times

Building programme
Date of completion
Whether the project has been discussed with the appropriate sponsoring authority
Whether the building is to be erected in one operation or by stages
Possible future extensions

Mechanical and electrical engineering services
Type of heating and fuel to be used; exposed or concealed form of heating
Lifts
Refrigeration

Table 1.2   **Information for building (continued)**

Air conditioning
Temperature and quantity of hot water
Type of lighting
Electric power requirements
Refuse disposal
Peak periods for service, and whether the building is to be used by night
Any special provisions

Finance
Amount of money available
Whether the building is a speculation or is for a specific use
Employment of a quantity surveyor, consultants and a clerk of works
Professional fees

General information

Space standards
Internal; external; offices; shops; garages and service stations; industrialised buildings; farm buildings; hospitals; restaurants and bars; public cloakrooms; theatres; cinemas; sports and swimming; educational buildings; libraries; hotels; housing.

Environment
Moisture; air movement; daylighting; heat; sound; thermal installation; electric lighting.

Utility services
Water supply; sanitary appliances; drainage installation; sewage disposal; refuse disposal; electricity; mechanical conveyors; firefighting equipment; ducted distribution of services.

Materials
Timber; boards and slabs; stones; ceramics; bricks and blocks; limes and cements; concretes; metals; asbestos products; bituminous products; glass; plastics and rubbers; adhesives; mortars for jointing; mastics and gaskets.

Construction
Structural behaviour; foundations; walls; framed structures; fireplaces; flues and chimneys; stairs; building operations and site preparation.

Components
Joinery; doors; windows; glazing; roof lights; ironmongery; balustrades; demountable partitions; suspended ceilings; industrialised system building.

Finishes
Flooring; plastering; rendering; wall tiling and mosaics; thin surface finishes; roofing.

In a project of any size it is usually the best policy for there to be one series of meetings specifically concerned with the management of the job and the various functions and disciplines involved. This may be known as the *"Planning Team Meeting"*, set up by the client's organisation to make working policies and to manage the job overall. These meetings should be chaired by the client's representative and attended by the leaders of all the design disciplines and the appropriate members of the client's own team.

This series of meetings is centred around the work of the design consultants and their interactions with the client's advisers, and it is thus not suitable for monitoring the progress of the job, and the needs of the various contributors to the design process. For this purpose a monthly *"Client Progress Meeting"* should be held between the design team leader (and any of his specialist consultants) and a representative of the client. It is generally advisable to hold these meetings in the design office, where personnel and information are readily available and it will seem natural for the designer to take the chair. Its prime aim is to report progress, ask for and give information and formulate working policies.

A further series of meetings is necessary to assist in the internal management of the design team's operations. These *"Project Team Progress Meetings"* should be chaired by the job leader and involve a progress report from each professional team leader, discussions on programme and working policies and requests for information. Whilst these meetings should act as briefing sessions for the monthly Client Progress Meetings, it may be found necessary to hold them more frequently at periods of high co-ordinating activity. However, their length should be kept as short as possible in the belief that the useful duration of a meeting rarely exceeds ninety minutes!

Graphic communication

The principal means of design communication is still graphical, though the nature of drawings and their presentation is tending to change with a concentration more upon the imparting of information than the meticulous draughting of every detail of construction. This change is most apparent in the introduction of standard details which rely upon a system of cross-referencing of information, the use of dimensional grids and ranges of preferred dimensions, and the introduction of computer aided design (CAD).

Once the detail design stage (E) has been reached, each design leader should prepare a *Production Drawing Programme* recording how many drawings are expected to be produced, when they should be completed and an estimate of how long will be needed to prepare the bill of quantities from the information they contain. Figure 1.3 illustrates such an overall programme.

Figure 1.3   **Project programme: stages E-H**

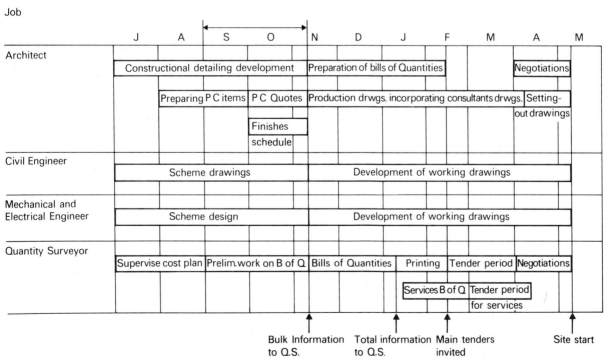

9

An example of a schedule of production drawings, identifying each piece of information to be produced by the architectural team, is shown in Table 1.3. The sequence of production and the likely duration of each item of work may be calculated from past experience and factors which influence production such as:

1. The quality of the product being designed

2. The availability of information sources

3. The content of work in each element

4. The multidisciplinary nature of the element being designed

5. The experience and quality of staff available.

The completed programme chart should be circulated to all the design leaders for their comment, suggested amendments and agreement. When a programme has been agreed for each discipline, an integrated master programme may be prepared against which progress can be recorded, and which indicates where information will be available and by whom it is to be used. Not all disciplines may agree to produce all their allocation of drawings during the formal production drawing period (stage F), and in these circumstances this fact must be noted on the programme and the effects assessed in the light of the flow of available information.

It is essential that a record is maintained of all drawing issues and a practice which has been found to work well is for a *Drawing Issue Schedule and Record Card* (Figure 1.4) to accompany each formal issue, recording:

(a) names of the sender and recipient

(b) drawing titles

(c) amendments made to date

(d) number of copies attached

(e) date of issue.

Figure 1.4 **Drawing issue record card**

Element

Sheet No.

Job No.

Job Title

No. of Prints

Additional   Prints
Negs

Architects Instruction No.   refers

Description   Drawing No.   Amendments

Distribution   No. of copies

Account
*Purpose
Date

Drawings issued on behalf of   by   *See Standard Key Symbols

These cards have the additional advantage of providing a record of actual drawing output performance, which can later be cross-checked with the weekly record sheets, which should be filled in by each member of staff, indicating how their time has been allocated between tasks.

As the communication function is so bound up with drawing practice, it is clearly desirable that there should be no confusion over the terminology used to designate drawing types and to define their purpose. British Standard 1192 contains recommendations for a standardised formal building drawing practice and the reader is advised to refer to this work (1) for a more detailed study of these recommendations, which can provide a valuable structure upon which a design manager can build his approach to drawing office practice.

It is convenient that a standard terminology should be adopted in order that some uniformity of understanding exists when reference is made to a group of drawings; it is important that the purpose and nature of drawings should be recognised by all those likely to use them. Whereas the group titles may vary, the British Standard suggests that the following should be used:

A. Design stage (A-D)

1. Sketch drawings
   The preliminary sketches or diagrams to show the designer's general intentions. These could incorporate the brief requirements agreed by the client and be distributed to the consultant design specialists to indicate the general nature and form of the project.

B. Production stage (E and F)

1. Location drawings

   (a) *Block plans.* To show the location of the site and the disposition of the building outlines in relation to the adjacent property and possibly to the town plan and development projects.

   (b) *Site plans.* To show the disposition of the buildings in relation to the setting out points, access roads, site features and overall layout; to include indication of service and drainage runs.

   (c) *General location drawings.* To indicate:
   (i) the positions occupied by the various spaces allocated in the building

   (ii) the general construction pattern

   (iii) the location of the principal elements, components and assembly details.

   These drawings may be dimensioned or related to co-ordinates in a dimensioned, or modular, grid.

2. Component drawings

   (a) *Ranges.* These give the basic sizes, the system of referencing and the performance data for a set of standard components; for example, a window or door frame units. These will form part of an office library of standard details, if one is maintained, from which appropriate sheets may be drawn for specific jobs and referred to on location and other drawings.

(b) *Details.* These should provide all the information necessary for the manufacture and application of the component. For example, they would show cross-sections through window and door frame units rather than overall frame sizes.

3. Assembly drawings

These comprise the constructional details of the building, particularly concentrating upon junctions both in and between elements and components, for example, window cill and lintol details showing the treatment at the interface of the frame and jambs.

4. Schedules

Much information can best be imparted in the form of a schedule rather than on a location drawing, provided that there is a good cross-referencing system to identify a unit in its location. Typical elements often best treated in schedule form would include:

(a) doors (with frames and ironmongery)

(b) windows (with frames and ironmongery)

(c) precast units — lintols, copings, etc

(d) finishings — texture, decoration, colour

(e) joinery fittings

(f) service installations — heating, electrical fittings, sanitary appliances

(g) floor finishings — material, colour, etc.

Wherever possible these schedules should have a direct reference to specifications and performance standards. In addition to their use as supplementary information for drawings, they can prove most useful for such functions as:

(a) measurement and taking off quantities

(b) estimating

(c) ordering materials and components and calling forward in accordance with programmed requirements

(d) checking deliveries

(e) forming basis for nominated contracts

(f) checking sub-contractors and suppliers

(g) locating items of work and progressing their completion.

C. Coding and annotation

If the aim of uniformity is to be achieved, there should be a standardised layout for drawings and identification of sheets. British Standard 1192 sets out recommendations for margin width for filing (minimum 20 mm), title panels, information panels and key indications of sequence of drawings. It also lays down a basis for drawing sizes and folding systems for filing.

There are several referencing systems for codification of information which have been favoured specifically by designers, and probably the most universally accepted is that known as 'SfB', although the UDC and CBC systems have their supporters. The principal requirements of a classification system for building are that it should:

(a) indicate the type of drawing

(b) define the element or component

(c) define the construction method

(d) designate the material

(e) be sufficiently simple to be memorable in the drawing office or on site without constant reference to an index.

The principle of classification of drawings and elements by SfB is to give information in as brief and unambiguous a form as possible whilst meeting the above requirements. Thus a reference could read as follows:

Type of detail	As	Assembly
Functional element	(22)	Wall
Construction and material	Nd4	Lapped sheeting, aluminium
Detail number	8	Place in the sequence.

The advantage of using a generally acknowledged classification system for information is that it can simplify referencing to other drawings or documents, use less space to impart information on a drawing, and make it easier to substitute new and improved details without requiring to redraw or amend an extensive number of drawings. In certain forms it will also fit easily into computerised systems of data storage and retrieval.

In this chapter it is not intended to take this subject further than to point out that the design manager should make himself aware of the need for such a system of classification which is suitable for application throughout the many functions of the design process, and it is recommended that the subject is pursued further through the references given at the end of this chapter.

Storage of drawings for easy recall is always a problem and several filing systems are available for the storage of design drawings and for their retention in easily accessible racks during the design period when they are required for frequent reference, by the designer at the drawing board. The use of microfile film is the subject of BS 3210, and has found its place in design filing systems. However, it can introduce difficulties where a designer may wish to refer to several drawings at the same time. It has obviously great value in cases where there is little available space for storing documentation and where there is a large volume of drawings to be stored.

The management function

If the design process is to be carried out effectively there must be a body of *management information* available to the manager. This will enable him to make informed major administrative decisions based upon facts rather than upon intuition or immediate expediency. The shared flow of such information can form the foundation of co-operative attitudes between those concerned in the project. It should assist in the identification of common goals and an understanding of the purpose underlying those decisions which are taken to achieve them.

If it is to be of real value, management data must have certain characteristics:

1. It must be available in a standardised form which is adhered to consistently across the range of topics.

2. There must therefore be compatibility between the different items of data both in the manner of presentation and classification, and in the contents themselves. For example, if metric measure is used, it must

be used throughout so that door and window sizes are compatible with their sub-frames or the dimensions of prepared openings; similarly measures of thermal conductance or insulation must relate to heat input and loss figures using related units.

3. The sources of the data must be identifiable and the range and frequency of its circulation generally known.

4. The cost of obtaining and updating the data must be established so that it can be justified in terms of its usage.

The control of the design process in a design office is normally concerned with the money which is available for running that office and which will form one of the major control parameters. In the case of an architectural practice, this money is usually obtained from the fees which are earned for the professional services offered and hence can be related to individual projects. Where the design function is a part of an overall construction service and the income from a project cannot easily be isolated, the money available to the design department will be allocated in the annual, or project, budget on the basis of the anticipated work load, staff employed and services required.

These are, therefore, important factors in the management of the design process, which must be orientated towards producing a design solution to a series of problems normally by an agreed time within a specified cost limit. To achieve this, the manager may be guided from a choice of data derived from sources both within the firm and from outside. It must be kept under review to ensure that it does not become outdated and hence misleading. Such data would cover such items as:

1. Manpower resources

   (a) Technical staff
       (i) Availability of those employed within the department
       (ii) Range of skills
       (iii) Range of salaries, probably classified by grades

   (b) Administrative staff
       (i) Number employed
       (ii) Experience and function
       (iii) Range of salaries

   (c) Ratio of technical to administrative staff, by number and by total salaries

   (d) Multi-professional team experience or mono-professional backgrounds; note of past formal or informal groupings in project teams

2. Money

   (a) Income breakdown, indicating the sources (fees or allocated budget) and timing

   (b) Expenditure
       (i) Direct. This includes salaries of technical staff (ie, design production
       (ii) Indirect. To include overheads relating to all aspects of administration, including salaries

   (c) Liquid assets. Cash, etc

(d) Reserves. Sums set aside for taxation, replacement of equipment, etc

(e) Fixed assets and investments. Premises, equipment, etc, and return on capital invested

Figure 1.5   **Distribution of staff**

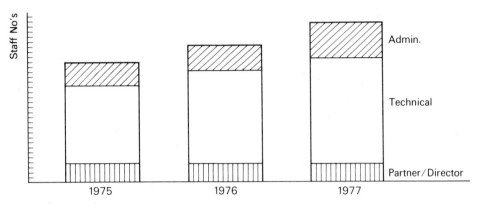

Figure 1.6   **Distribution of income**

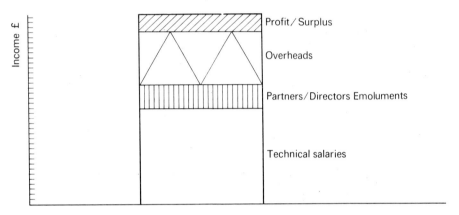

3. Accommodation and facilities

   (a) Space available for different types of working – specialised project groups, etc

   (b) Special draughting or computing equipment available or on hire

4. Productivity ratios

   There are many ratios which can be used by management to indicate the financial viability of an enterprise's activity. In selecting those most appropriate for a particular situation it is necessary to remember that no office can continue to function satisfactorily when its expenses exceed its income and such ratios can help to show where expenditure may be occurring at an unduly high rate when compared to past experience on other projects. Such ratios are normally related to production costs; in this case the cost of producing design solutions and communicating them to others. Those which have been found to be the most indicative are:

   (a) Income per annum (fees earned or allocated funds) per head of technical staff

   (b) Income per annum per £1 of technical salary

   (c) Total costs per head of technical staff

(d) Average salary per head (by grades)

(e) Net profit as a percentage of turnover

(f) Direct expenses as a percentage of turnover

(g) Overhead costs as a percentage of turnover.

Figures 1.5 and 1.6 illustrate graphical means whereby some of the information from these ratios may be presented.

Programming

If the design process is to be controlled it is necessary to estimate the resources required to execute the project and the periods for which they are likely to be employed. This refers closely to the points just made as it is necessary to obtain some assessment of the work involved in a project, the skills demanded and the overall duration allowable for each stage.

It is generally agreed that it is very difficult to "design to order" and therefore any attempt to predict the manhours required to accomplish each stage of a design must be doomed to failure from the outset, at least in the mind of the traditional designer. Indeed, this is the attitude which has so often resulted in delayed starts, or in work commencing without adequate information.

However, it must be admitted that situations do arise where unforeseeable problems may be encountered in the development, perhaps, of an unusual solution in planning or construction. Nevertheless, these are comparatively rare, and there is nothing like the setting of production objectives to avoid the open-ended attitudes encouraged by Parkinson's Law. Ultimately, under any system, there must be a point by which all the decisions in a stage should be made, and the earlier this time limit is defined and accepted by all those concerned, the less chance there is of ill-considered last minute decisions being made.

The objective of a programme is to provide a statement of intent, a "master plan", against which actual performance may be measured. With the establishment of planning standards based upon known levels of performance and budget allocations, it is possible to carry out management planning in financial terms for both the office as a whole and for individual projects.

If records are kept, however imprecisely, on a regular basis, of the time spent by members of staff on each stage of a project, the feedback of such data will eventually provide a body of knowledge from which some estimate of workloads can be made and variations from the anticipated performance will need to be accounted for. The data will, of course, include variables. No two designers necessarily work at the same speed and their creative patterns or working may well be very different. Design problems can make widely ranging demands upon the designer in view of the uniqueness or normality of the problem, its complexity or its interdisciplinary nature.

The *financial planning* of the project or of the performance of the office as a whole must be based upon a comparison of income with the following information:

(a) average total cost per man day

(b) the desired level of profitability (or excess of income over expenditure)

(c) the planned expenditure broken down into each stage of the work so that financial check points can be planned to monitor progress.

In the event of the work being carried out for an agreed fee, (for example, the RIBA fee structure), there will probably be some system of stage payments which should be reflected in the planned expenditure programme. Too often an excessive amount of expense is incurred in the earlier stages of work, due to taking longer than had been planned on the initial design and/ or production drawing stages, or to employing more expensive staff than anticipated, so that insufficient funds are available for adequate commitment to the preparation of the production drawings or to the supervision of the production stage without encroaching upon the profit margin which had been intended.

Figure 1.7 illustrates a labour expenditure control chart based upon the stages of the RIBA Plan of Work. This may be used as the master control chart with more detailed versions being prepared to cover each of the stages and thus provide a more valuable means of checking performance in sufficient time to take some corrective action, ensuring that the desired position has been restored by the end of the project.

Figure 1.7   **Labour expenditure control**

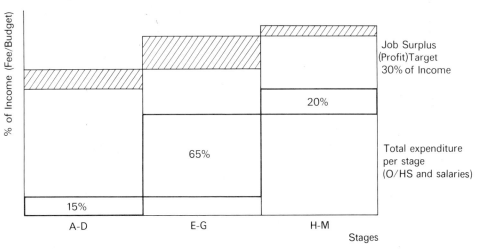

A combined chart for the workload of the whole office can be compiled from the accumulated information shown from these charts for individual projects current in the office. From this can be gained some indication of the anticipated cash flow for the foreseeable future period, though a more detailed analysis of this aspect would need to be carried out by the accountant to detect any danger of over-commitment during any specific accounting period. Such factors as bank loans, overdrafts and special payments to staff would need to be brought into such a calculation, together with the effects of inflation upon monetary values.

After these control charts have been drawn, and to some degree parallel to their preparation, a staff commitment programme as illustrated in Figure 1.8 must be drawn to show the staff of each salary grade, their planned work programme and the periods when they will be freee to undertake new work. Such a chart is invaluable when allocating staff to a new project and matching the cost of salaries with the anticipated expenditure, whilst forming a well-balanced team incorporating the right blend of expertise and experience at the appropriate stages of the work. The role of the partner in architectural practice, or the departmental head or director in other cases, is very important when carrying out this calculation as his participation even in a part-time capacity may well be critical and must be charged as a proportional amount of his salary or emolument.

17

Figure 1.8

Figure 1.8 Staff commitment programme

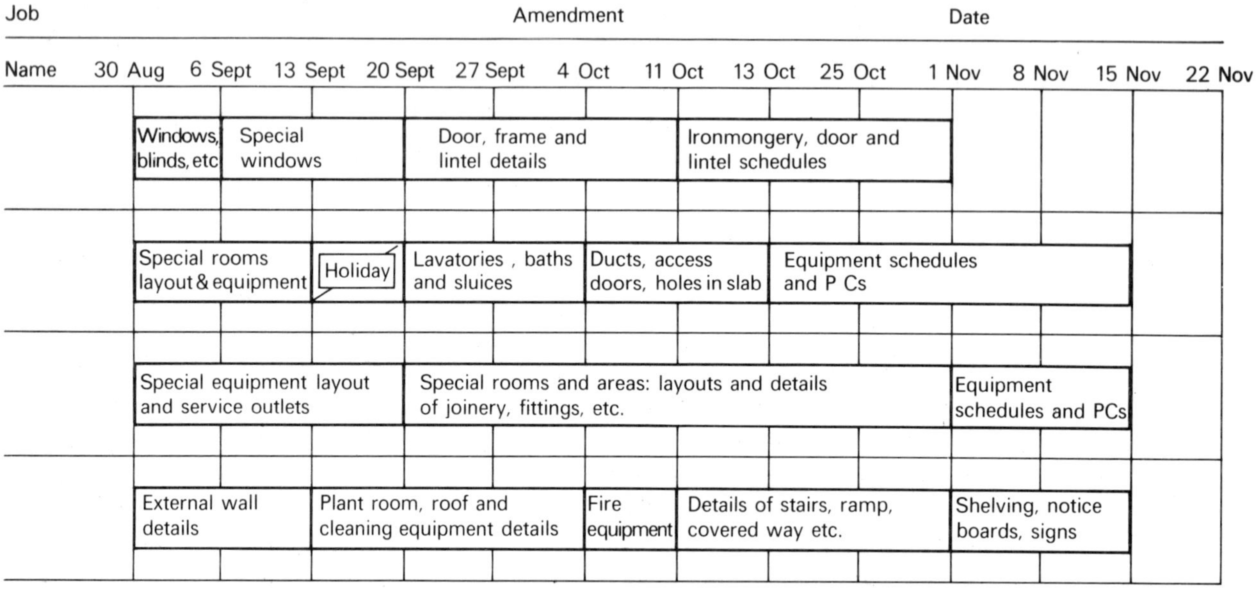

Job						Amendment					Date		
Name	30 Aug	6 Sept	13 Sept	20 Sept	27 Sept	4 Oct	11 Oct	13 Oct	25 Oct	1 Nov	8 Nov	15 Nov	22 Nov

Windows, blinds, etc | Special windows | Door, frame and lintel details | Ironmongery, door and lintel schedules

Special rooms layout & equipment | Holiday | Lavatories, baths and sluices | Ducts, access doors, holes in slab | Equipment schedules and P Cs

Special equipment layout and service outlets | Special rooms and areas: layouts and details of joinery, fittings, etc. | Equipment schedules and PCs

External wall details | Plant room, roof and cleaning equipment details | Fire equipment | Details of stairs, ramp, covered way etc. | Shelving, notice boards, signs

As every design project has its own particular problems, demanding a level of skill and experience, care must be taken to identify these requirements in the planning stages so that adequate time and skill can be allocated to their solution. Where there is a discrepancy between the cost of these staffing decisions and the sum allocated for that stage, preference should be given to the quality of staff to be employed, if after re-examination the decisions still appear wise. A note of this point of probable over-spending should be made and the danger borne in mind when the work is being done so that any acceptable economies may be introduced, or the deficit noted and attempts made to recover it later, always provided that any action is not taken to the detriment of the job.

This all means that wherever possible the grade of staff must match the work to be done. In building terms, it is uneconomical to employ skilled tradesmen upon unskilled work except in an emergency. The use of a library of *standard details* with a cross-referencing system can save considerable expense and time in work of a repetitive nature. It also has the advantage of enabling design time to be devoted either to the solution of new problems or to the refinement of details whose performance has already been tried and recorded on other projects. By the issuing of such detail drawings the need arises for a comprehensive system of classification and referencing which can be used to identify the location of these details on the layout drawings or on the schedules. The manner by which design information is transferred to other parties is discussed elsewhere in this chapter.

It must therefore be accepted that, in spite of the variables likely to be encountered, design production norms can be established both for building types and individual members of staff if adequate records are conscientiously maintained. From these a reasonably reliable estimate can be formed of the manhours required to complete each stage and hence of the manpower needed to carry out the work within the planned schedule, taking into account the possible delays likely to be encountered whilst awaiting approvals or instructions. The analysis of the project itself has already been discussed earlier in this chapter. If salary grades are then applied to the staff allocated to the work, it should prove possible to gain an indication, to some ±5-10% accuracy, of the likely cost involvement of the project.

Examples, already referred to, of such staff programmes are shown in Figures 1.5 and 1.6. Figure 1.7 includes all members of the design team who are engaged in a stage, whilst Figure 1.8 includes only one design office's programme. Except in the cases of multi-professional offices it will not be possible to provide a costing for more than one profession's work.

We have used the RIBA Plan of Work as the basis for the design process in building as it is geared to the generally accepted procedure adopted by most clients, designers and contractors, and the reader is referred to this for a detailed examination of its function through all the stages of the design. However, it has one major weakness in that it fails to emphasise the reiterative nature of the design process, which has already been mentioned, and gives the impression of a sequential pattern with little opportunity for back-reference.

Other methods have been developed to introduce greater flexibility into the design activity and greater interchange between the designers involved in a project, to enable decisions to be taken at as late a point as practicable. One method specifically evolved for use in architecture and building design in general, as an alternative to the Plan of Work, is CASA, or the collaborative strategy for adaptable architecture. The aim of this method is to enable everyone concerned with the designing of a building to influence decisions that effect both the adaptability of the building and the compatibility of its components. The chief feature of this system is that the main structure of the building is designed first, which will permit a certain range of options in layout and component design by determining the structural system to be used and the overall dimensions of the layout, number of storeys, service cores, etc. A contract is then let for this part of the building and the work of construction begins. While this is proceeding, the layout is planned in detail within the options permitted by the structure and the contract for this work is let in turn. By this means this sub-system can later be remodelled to meet any new requirements resulting from change of ownership or use in the future.

The CASA system is an attempt to avoid the restricting effects of taking major decisions about structure, services and methods of construction at an early stage in the design. Normally the building form is determined and fixed before many aspects of the design of the structure have been explored and many alternative solutions considered. It also has the advantage of leaving the layout design to a later stage, giving more time to establish the needs of particular users without the penalty of extending the total design and construction time.

Unfortunately the method requires a re-organisation of the pattern upon which the traditional building process normally works. It demands an equal participation in the major decisions related to thebuilt form from both architects, engineers, contractors and quantity surveyors. It also expects that the clients must be willing to release money for the construction of a building, the precise detail of which is at that stage not fully known, but could be one of a range of predetermined options. These changes will not be accepted easily, although there are signs that practice is moving in this direction: for example, the internal layout design of office buildings is now frequently decided only when the needs of the tenants are known, and several building systems are available which permit a variety of changes in cladding and interior organisation within the lifespan of the building. However, the sequence of decisions will need to be reorientated if this added flexibility is to be gained universally in the design of buildings.

Table 1.3   **Drawing list (architect)**

Element	Drawing title	Scale	Number
(0-)	*Total building*		
	Plans - 'A' level (underpass)	1:50	1
	'B' level (plant, stores, etc)	1:50	2
	'C' level (library)	1:50	1
	'D' level (entrance, common room,		
	lecture theatre)	1:50	2
	'E' level	1:50	2
	'F' level	1:50	2
	'G' level	1:50	2
	'H' level	1:50	2
	'J' level (animals)	1:50	2
	'K' level (roof plant)	1:50	2
	Key elevations	1:100	4
	Key sections	1:100	2
(1-)	*Substructure*		
	Various architectural details, including underpass	Various	2
(2-)	*Superstructure*		
(21)	Details external cladding - main block		
	- library	1:5	1
(22)	Details internal partitions	1:20 & 1:5	2
	Details toilet and shower partitions	1:20 & 1:5	2
	Details changing rooms, etc, animal floor	1:20 & 1:5	1
	Dog pens	1:20 & 1:5	1
	Room dividers	1:20 & 1:5	1
	Room elevations and associated details for		
	library, lecture theatre, etc	1:50	5
(24)	Stairs - section and plan	1:20	1
	- details	1:5	1
	- access stairs to roof plant room	1:20 & 1:5	1
(27)	Roof - plans - main block	1:50	2
	- library	1:50	1
	- section through main roof and plant room	1:50	1
	- details - main block	1:20 & 1:5	2
	- library	1:20 & 1:5	1
	- entrance canopies	1:20 & 1:5	1
(3-)	*Secondary elements*		
(31)	External doors - main block	1:20 & ½FS	2
	External doors and windows - library	1:20 & ½FS	1
	Window types and details - main block	1:20 & ½FS	1
	Louvres openings	1:20 & ½FS	1
(32)	Internal doors - types	1:50	1
	- details	½FS	2
	Glazed openings and hatches	1:20 & ½FS	1
	Duct covers and hose reel recesses, etc	1:20 & ½FS	2
(35)	Ceiling co-ordination plans (based on (0-))	1:50	16
	Ceiling details - general	½FS	1
	- lecture theatre	½FS	1
	- library	½FS	1
(4-)	*Finishes*		
	General details	Various	1

Table 1.3   **Drawing list (architect) (continued)**

Element	Drawing title	Scale	Number
(5-)	*Building services*		
	Builders work	Various	4
(6-)	*Building installations*		
	Builders work	Various	4
	Lifts	Various	4
(7-)	*Fixtures and fittings*		
(72)	General fixtures -		
	Service spines	1:20 & ½FS	1
	Worktop and sinktops	1:20 & ½FS	3
	Library counter	1:20 & ½FS	1
	Audio visual carrels	1:20 & ½FS	1
	Lecture theatre - teaching wall	1:20 & ½FS	1
	- console and seats	1:20 & ½FS	1
	Fume cupboards	1:20 & ½FS	1
	Balance benches	1:20 & ½FS	1
	Ventilated gas cupboards	1:20 & ½FS	1
	Darkroom wet benches	1:20 & ½FS	1
	Dunk tank (animal floor)	1:20 & ½FS	1
(77)	Special fixtures -		
	Autoclaves	Various	1
	Incinerator and cage wash	Various	1
	Tea bar counter	Various	1
	Glazed extract hoods	Various	1
(9-)	*External works*		

TOTAL   116 A1 size sheets

BIBLIOGRAPHY

1 B S 1192 (metric) 1969. Recommendations for a standardised format for building drawing practice.
2 CROSS, N. The automated architect. Pion, 1977.
3 ASIMOW, M. Introduction to design. Prentice-Hall, 1962.
4 BROADBENT, G. Design architecture. Wiley, 1973.
5 RIBA. Handbook of architectural practice and management. RIBA, 1965.
6 RIBA. Plan of work. RIBA, 1967.
7 DRUCKER, P F. The practice of management. Heinemann, 1955.
8 JONES, J C et al. Conference on design methods. Pergamon, 1963.
9 TAVISTOCK INSTITUTE. Interdependence and uncertainty: a study of the building industry. Tavistock Publishers, 1966.

2.The Elements of Production

"Production is any process or procedure designed to transform a set of input elements into a specified set of output elements."

M K Starr

The construction industry has many facets which are far removed in their nature from those of the generally acknowledged manufacturing industries. For example, the production of tinned soup, lengths of terylene and cotton fabric or a washing machine may seem at first thought to have very little in common with the construction of a hospital, a school or a housing estate.

However, if the processes which are involved are broken down and their basic elements are carefully analysed, it will be found that there are fewer divergencies than may at first seem apparent. Anyone concerned with production methods and control in any industry should therefore start his studies with a review of the principles of production planning and control as applied to the greater body of manufacturing industry where the subject has received the greatest attention for the longest period of time and has been the centre of applied analytical thought. It will readily be seen that there are many areas of common interest which are fundamental to a manager's approach to his task whatever product may be his prime concern.

The differences tend to lie in the conditions under which the production is carried out and in the facts which control both the nature of the product and the factors which stimulate demand.

The nature of the product

The characteristics of the product which is being manufactured naturally exert an overriding influence upon the production processes and hence upon the control procedures which are most appropriate and the problems which face the manager.

Products are basically of two types:

(a) they are produced wholly within the compass of the factory in that they are made from an input of raw material which is transformed within the manufacturing process into marketable products for despatch to the consumer directly or through established outlets; sugar and fertilisers are examples of this type.

(b) they are the result of the bringing together of wholly or partially prefabricated components which are assembled in the factory and rely in part upon sub-contractors and specialist suppliers to make their contributions to the process; a motor vehicle and a television set are examples of this type.

A building as a product falls into the second category but has other characteristics which distinguish it from other manufactured products, not least those which influence the conditions under which the production is carried out.

Factory production lines are carefully worked out to give the most advantageous flow patterns for a product to move through the factory from the earliest stages to the despatch bay whence it is finally transported away from the factory, or is transferred to storage where it can form part of a buffer stock against unpredictable fluctuations in future market demand. Thus the principle upon which the manufacturing system is based is one of routing the product through a sequence of operational stages at each of which its form is transformed in some manner, either in its profile or its intrinsic properties, or by having other components added to it. The factory can therefore be located conveniently for its supply routes for raw materials or component deliveries, or for onward despatch of its output, and once the plant layout has been designed it can be established for as long as the product is likely to be marketable, probably with only minor modifications to the machinery. This also means that the workplace at each stage can be laid out to facilitate the manufacturing processes with full control of the lighting, temperature, humidity and other environmental conditions, as well as check points for quality control.

The building product is essentially different in that in its final constructed form it is by nature static and the factory rather than the product moves on. It is thus not possible to design a permanent production layout which can be used consecutively for a series of outputs. Whilst it may be possible to fabricate components under normal factory conditions (and this is becoming increasingly the case) the actual assembly processes must take place in the location in which the finished product will be used, however inconvenient this may be in regard to the transportation of resources, both men and materials, and however unsuited the microclimate may be for construction work. Problems of establishing quality check points to monitor standards are liable to be accentuated so that normal industrial quality control methods are rarely applicable.

It is therefore a particular aspect of the construction manager's expertise that he must be capable of adapting to a wide range of conditions under which he must carry out his work effectively. Except in the case of housing contracts it is very unlikely that he will have encountered a similar production task in all its facets. The design is unlikely to be wholly standard, the labour force will probably be newly recruited at least in part, the location of the site will be untried and the conglomerate of materials and components which are specified will have a fair proportion of items not previously encountered, together with a number of specialist sub-contractors whose level of performance will be an unknown quantity.

The nature of the product is also complex, demanding a wide range of skills, both technical and organisational, and in essence it incorporates heavy materials and the use of both wet and dry processes and in many cases defects are easily covered up and only become apparent as a result of use over a lengthy period of time. A building is not a product of a throw-away philosophy as the financial investment which it represents anticipates a life of at least thirty years and in most cases one of nearer sixty; thus it must be produced to perform well in use over a considerable period.

The nature of manufacturing processes

Production is generally understood to mean any sequence of operations which transform materials from a given to a desired form. In construction, as in many other industries, it tends also to embrace those assembly processes which bring previously manufactured components together to form a new product such as a building, a television set or a ship. Thus it is usually accepted that the transformation of materials into products may incorporate more than one stage of operations; indeed, it may involve one or a combination of the following:

1. Disintegration

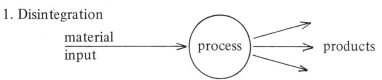

2. Integration and assembly

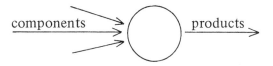

3. Service

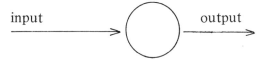

Disintegration

This is the process whereby several products are obtained from a single raw material source which is subjected to some form of transformation. Examples of this process are the breaking up of logs in a sawmill to make a variety of timber products, or the melting down of ingots to cast steel bars, or the conversion of clay into ceramic products through firing. The essential element of this process lies in the establishment of the most fitting method of transforming the raw materials into their new states, since such a process must be economical in both cost and material resources, and, in the light of today's energy crises, restrained in the use of energy in its conversion processes. Many well accepted materials for building products (copper and aluminium are prime examples) are very costly in terms of the energy which they consume in their processing, whilst natural materials, such as those in the cellulose group, require very little energy. It would seem that the consumption of energy will become a prime factor in deciding where the emphasis of research and development may lie in the future.

Integration and assembly

These are the production processes most frequently encountered in the building industry, where components fabricated off the site are brought together and assembled, with or without the addition of site-prepared materials such as mortar, concrete or adhesives, in order to produce the completed building. Viewed in this way the building process can be seen to reflect on a much larger scale the assembly of a complex piece of machinery or a range of units for telecommunications.

The principal elements of the process lie in a well organised system of ordering and delivery of the components and a rigorous attention to dimensional tolerances and assembly jointing details which are compatible with the circumstances under which they will be carried out.

Service

These processes are concerned with improving the state of a product, perhaps by toughening, watersealing or painting, or simply by carrying out maintenance where performance has been impaired through use. Many public utilities are service industries in that transport companies provide a means of changing the location of persons or products as a part of some other overall process, such as getting to the office or distributing the mail. Hospital services carry out a form of repair and maintenance in the field of health, whilst the building industry itself may be said to provide service in the form of buildings within which useful activities may be performed. Its role in the maintenance and rehabilitation of property is clear as the need for adequate accommodation increases in a period of rising living standards. This issue, always an important one, has grown in significance now that the advantages of conservation have become apparent in a time of world shortages which have demanded the demise of the throw-away philosophy which has been adopted by much of the consumer product industry in place of high standards of maintenance service to ensure the efficient performance of a product whether it is a building or a washing machine over an acceptably long lifespan.

Figure 2.1   **The production cycle**

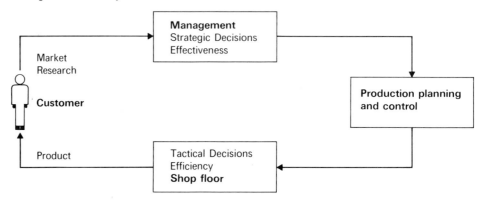

MANUFACTURING SYSTEMS

Much will be written in later chapters on the role and responsibilities of the production manager in the construction industry and in these sections detailed studies will be made of his duties and the techniques and strategies which are open to him. However, any discussion of the elements of production would not be complete without some reference to production management's responsibilities in the execution and control of production processes in industry as a whole, as the pattern which emerges is of universal application and defines the objectives of the product process as a whole. Production management is concerned with two responsibilities at two different levels:

1. The design of the production system

This is a long-term strategic responsibility involving non-repetitive decisions which, once they are taken, are virtually irreversible and unique to a specific project. These decisions lay down the basic pattern of production for a factory, defining the range of products to be manufactured, the principles of production to be followed and the anticipated rate of output needed to match demand. It is particularly important in the construction industry, where the large number of "one-off" or unique contracts is very high and strategic decisions affecting the whole life of an individual product are encountered more frequently than in other industries. Whereas this may be said to increase the decision-making load of the production manager in construction, there is the compensating advantage to be gained from the fact that the construction contract is comparatively short-term and thus there is opportunity for rethinking strategy at more frequent intervals than in the case of most manufacturing industries.

2. The design of the control system

This is a much shorter-term responsibility taking advantage of any elements of repetition which may exist in the project and which will provide some basis for feedback data on production performance. The control system is thus of a tactical nature encompassing the many production centres together with the advisory or service departments which are involved. It is therefore important that, in its various parts, it should be capable of modification to allow for adaptation to the variety of production processes being used.

The major problem inherent in production management is frequently the difficulty of reconciling the needs of the strategic decisions concerned with the production system with the tactical control decisions which are concerned with the implementation of the overall plan and the actual methods by which the production is carried out.

Three factors determine the place of production planning and control in an organisation:

1. The type of production.

2. The size of the factory or manufacturing centre.

3. The characteristics of the industry.

It is desirable that the nature of construction should be reviewed in the context of other industries, as we have already stated, so that a better understanding can be gained of the characteristics which are specifically attributable to it and which influence the way in which production can be carried out and controlled.

Types of production   There are three basic groups into which types of production processes may be divided:

1. Job

This is the "one-off" product which may be repeated at intervals but generally comprises only a small quantity of similar units involving little repetition. It is the typical case met with in building projects. Indeed, it is rarely (usually only in the case of housing) that a building is repeated in all its processes. Variations usually occur at least in ground conditions, and frequently in the juxtaposition of individual units of construction, whilst the actual location of the place where the production or assembly processes take place is constantly varying and can prove most influential. For example, the construction of identical floor slabs at ground and fifth floor levels can provide very different problems for production control. Whereas detailed operations within a sequence may be repetitive, the whole sequence is often unique.

2. Batch

This is used where a number of identical products are manufactured in response to a specific order or a known continuous demand over a period of time. The size of the batch and the frequency of its repetition are determined by the known demand for the product which governs the output schedule. It has universal application and examples may be seen in the production of pre-cast concrete units for a major contract or in another sense in the sequence of trade operations in a housing development moving through a predetermined grouping of dwelling types. The important characteristic of this type of production is that it is control-orientated as care needs to be taken constantly to maintain the balance of resources to provide the rate of output. Through the elements of repetition in each batch it affords the opportunity for feedback of performance information and thus for corrective action if it is necessary before the next batch is processed.

3. Continuous

This type of production is designed to meet a high rate of demand sustaining the justification for continuous output. Generally it is based upon the engineering skills necessary to design the production flow through the processes needed to maintain continuity and avoid the stopping of any part through an imbalance of process times or quantities. It is used typically in chemical processing and the manufacture of sheet glass, which is a continuous operation relying upon the maintenance of high temperatures which can only be built up gradually and which are a major item of expenditure once they are allowed to cool down.

The phrase *mass production* is used to describe the type of continuous process where a large number of identical articles are produced but where the machinery is flexible to the extent that no major modifications other than retooling are necessary to change the form of the product. This is the case in sheet metal press shops and to a smaller extent in precast concrete factories where a number of standard products may be made entailing only slight adjustment to the shape of the moulds. As projects increase in scale, so the opportunity for mass production techniques is enhanced. For example, there are instances of large quantities of windows being manufactured to a specific design for a particular development, and similarly in the case of doors and furnishings.

The phrase *flow process* is applied to the process which is set up for the manufacture of a particular and highly specialised product, such as an automobile engine or a chemical process. It has very little flexibility and once there is a change in the product the whole production line usually requires to be redesigned.

In many aspects the control of a continuous process is simpler than in the case of either job or batch production, but to be effective it must be planned in great detail to ensure that there are no bottle-necks created by variations in the rate of production of different stages of the sequence. In the case of building contracts it is usually possible to adjust the rate of progress of individual operations by the balancing of gang sizes or the number of machines employed, or sometimes even by reversing a sequence of operations. In a truly continuous process it is obvious that any imbalance could prove disruptive to the overall process. Indeed, this fact has not gone unnoticed by groups wishing to exert pressure in the event of industrial dispute where action by a carefully selected key group, however small, can quickly stop the whole production line.

Size of the factory

In the construction industry the workplace is the site, and as such it is exposed to all the uncertainties mentioned in Chapter 6. However, it must not be forgotten that the site constitutes a temporary factory and is a place of manufacture and assembly no less than in the case of its more permanent industrial counterparts. It may be felt in terms of size that the construction site employs few operatives compared to other industries, but it must be recalled that in all industry some 98% of factories employ fewer than 500, so that the site is by no means the exception.

The larger the factory, the more formally constituted must be the production control and planning procedures and the more clearly defined should be its place in the firm's organisation and its lines of communication. The smaller the factory, the greater is the reliance likely to be upon informal verbal communication links and less well defined procedures, decisions being made as the needs arise. One additional problem which faces the construction manager is the geographical separation of his head office from his sites, with increased communication links and less direct contact than is the case in other manufacturing industries. This usually means that there is a greater degree of responsibility and authority in site management to determine the production system.

Type of industry   It has already been shown that examples of each type of production occur in the construction industry, though there is a predominance of the 'one-off' type. Indeed, job production tends to be found most frequently in engineering based industries, whilst continuous processes are usually to be found in chemical processing, as has already been noted. Batch production, on the other hand, occurs in nearly all industries. Problems, however, vary between industries and are closely related to the nature of the product and this is discussed in another section. However, an illustration of this could be the influence of the materials which comprise the product upon the organisation of its manufacture. The problems facing a good processing plant differ greatly from those of a colliery, due to the perishable nature of the constituents on the one hand, and the durable imperishable nature of coal on the other. These influence storage facilities, scheduling of the raw materials, policy on stock holding and rate of turnover, as well as transport methods, to mention but a few.

Construction also has its share of perishable materials. For example, there is a limit to the time which may elapse between concrete being mixed and placed, which can become a governing factor in the design of the ready-mix delivery truck and even in its method of placing. Other materials deteriorate when inadequately stored and left unprotected from the elements.

The nature of production resources

The production cycle for any manufactured article starts and ends with the customer whose needs are to be satisfied as shown in Figure 2.1. A simplified version of this cycle is illustrated in Figure 2.2, which does not show the many sub-divisons into which each heading could be further extended. For example, a fuller illustration would include such functions as sales, budgeting, inspection and material procurement, all of which are discussed in the later chapters, and we shall be content here to review the objectives of the cycle and the resources which are used to provide the means of their achievement.

The prime objective is naturally to produce in accordance with the firm's interests. The strategic decisions of management relating to product development and design and sales forecasting, which are normally part of the pre-planning functions, are not often to be found within the control of the construction management team as they have already been made before the contractor is appointed. Exceptions to this are now on the increase and, besides the obvious case of speculative housing, there is a growing interest in forms of contract which give the producer a greater say in these aspects, through arrangements of negotiation and 'design and build'.

The design specification

The design specification is thus usually already established though its interpretation may still be a matter to be resolved in the light of the product's anticipated characteristics. So there are also subsidiary production objectives related to the components and assemblies which are to be incorporated. These, in turn, will require design specifications and also clear definitions of their functional performance together with their dimensional requirements of fit and interchangeability.

Figure 2.2   **Organisation of production cycle**

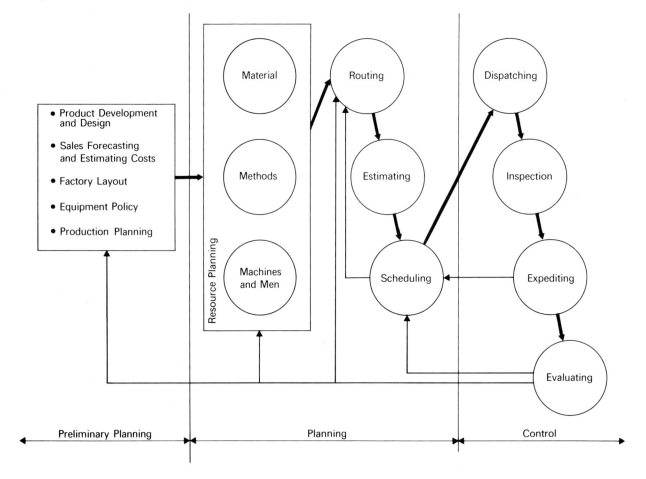

The means by which these objectives are to be achieved are the resources upon which production management has to call in its planning. Basically there are four production resources (the 4 M's), to which a further two may be added without which the picture is incomplete. The prime four are:

1. Materials

2. Methods

3. Machines

4. Manpower

to which should be added Management and Money.

Materials

Unless the materials and components are available at the work face in the right quantity and quality and at the right times the production process cannot meet its schedule, and to ensure that they are available must be a primary function of any production planning procedure. Unlike many other industries the range of materials or components which may be specified can vary widely between contracts and it is only with the introduction of the use of performance specifications allied to those in manufacturing industry that the producer is gaining more control over those which he may select. Thus the advantages of such tried industrial principles as standardisation and simplification are becoming recognised as applicable in the building scene, though the nature of the product still motivates against its general acceptance. Production largely depends upon the specification (dimensions, tolerances and quality), the accuracy of the inventories of required materials, their availability and the procedures of progressing their procurement.

Methods

The overall method of construction is determined by the product design, whilst the definition of methods for the execution of the work, the sequencing of operations and the selection of the processes to be used for specific operations, plant selection and movement patterns for men and materials are functions of the production management. The allocation of responsibility and the setting up of production centres form a part of the planning procedure. Control is exercised by establishing an agreed method statement and by the evaluation of the methods by judging their efficiency.

Machines

Mechanisation is increasingly replacing manpower on sites as it can handle more easily the greater loads required by modern techniques and packaging systems, and also because the replacement of manpower by machines reduces possible problems of man management at a time when labour costs are rising without any assurance of the output levels likely to be expected. Thus any production planning procedure must take into consideration all available data on plant specification, capacity, availability and reliability. It will usually interrelate plant and manpower in its production flow to ensure an optimum balance between these two resources, analysing the nature of each process to determine which is the controlling factor in each case. Control will be maintained by keeping a close watch on delays, breakdowns and levels of plant utilisation, together with an analysis of the effectiveness of plant maintenance policies.

Manpower

Every product requires a different mixture of skills and the effectiveness of any manpower resource is conditioned by the size of the force and the quality of the skills available for recruitment. The levels of its productivity and utilisation are dependent upon such factors as the working conditions, the planning operations and demarcation of responsibility, the working methods, and the degree of motivation provided by financial and other incentives. The changing technologies of construction have led to a redistribution of demand for individual skills and consequently to problems of recruitment, which are not helped by the minimal degree of repetition in the nature of the work so that newly won experience can easily be built upon.

Thus production planning procedures must be based upon the availability of labour resources rather than upon the opportunity to train, or retrain, personnel in specific skills, and it must adjust its output schedules accordingly. Production control is concerned with the monitoring of output, in particular the analysis of data relating to compliance with the planned schedule of work, rationalisation of tasks, and the effectiveness of incentives. It is a common complaint of management in construction that feedback of information on labour performance is unreliable, and therefore of doubtful value, when used as a control parameter. In comparison to other industries this is probably true but nevertheless the best available of such information must be used if any semblance of labour production control is to be initiated.

Management

The effectiveness of the systems of communication and control are dependent upon the skills and experience of management and its ability to adapt to changing situations, to recognise signals of danger or opportunity, and to provide leadership and direction. Control is maintained by the evaluation of results, which reflect the effectiveness of the procedures which management has installed. In construction the nature of the work often demands flexibility in decision making and there must be willingness to accept that communication and control systems may need to be redesigned if they are to be effective for a variety of contracts.

Money

All production requires financial resources to underwrite its activities, though the amount of capital involvement in construction tends to be very much less than in other industries where high investment in plant is often necessary. This limited amount of monetary resources makes it equally, if not more, important to maintain a rigorous financial control over such aspects as cash flow, liquidity and the allocation of resources to individual fields of the firm's activities.

After the above brief comments on the production cycle and the resources necessary for its operation, it is important that the reader should bear in mind that there are four factors which form the main influence upon any system of production which may be designed. These four factors are:

1. *The quantity or volume* of production which is required in the light of known or anticipated demand and the range of products likely to be manufactured or assembled in the one factory or site.

2. *The quality standards* demanded by the consumer and upon which a production control system must be founded and the level of skills determined.

3. *The time* for which the production system is designed to cater, both with reference to the rate of production for individual products and also in the light of the overall duration of the demand for the product. In the case of a building project this is normally determined by the contract period from commencement of work on site to handing over on completion, but in the case of speculative housing, for example, the rate of production may well depend upon an amalgam of the rate of sales and the projected increase in house prices. It must be recalled that within broad limits it is usually possible to vary the rate of production by increasing or decreasing the resources. However, by careful analysis it is possible to determine the rate which is the optimum to meet the limitations of either productive resources, cost or time.

4. *The price* for which the product can be marketed, which will govern, and to some extent be governed by, the other three factors, as it is closely linked with the resources which are available or need to be available for manufacture.

It is always important that management should differentiate between the price of a product to the consumer and the cost of its manufacture, which will include allowances for overheads and profit. Certain products have a

32

fixed price set upon them before manufacture commences and in construction the fixed-price contract was almost universally accepted until inflationary conditions made it impracticable to undertake work which was not covered by some clause allowing for some adjustment in price. In the case of a manufacturer producing a new product for the public market, it is essential that he must evaluate the anticipated production costs for various quantities and periods of output before he can decide how feasible a proposal may be, as beyond a certain limit an increase in the price of a product will result in the law of diminishing return. This is a situation not so frequently encountered in the construction industry except in periods of shortage of work and highly competitive tendering, which may result in such a lowering of tender prices that the narrowness of the potential margin may not be considered worth the risk of financial loss.

Assessing the demand

Demand for buildings arises from those who wish to occupy them, or to use them commercially as vehicles for investment, or to provide them as shelter for a public service. The custom for the industry's products derives approximately equally from public and private sources, though in the ultimate it is the government's financial policies which control both sectors and thus the level of demand.

So much has been written about the role of the construction industry as a regulator of private and public investment that there is no cause to add to it here, except to remind the reader how dependent the industry is upon the availability of financial resources such as mortgage loans and the level of the minimum lending rate on the one hand, and the fullness of the public purse on the other.

The effects of curtailment or expansion are felt over a long period of time as there is a considerable time lag, usually measurable in years, between a commission entering the designer's office and work commencing on site. The process of production is also lengthy, so that there is always some buffer of work in the pipeline. This means that fluctuations are not felt immediately and the repercussions are slow to take effect, which adds to the difficulty in forecasting demand. A further complication is that completed designs may well be shelved or brought forward at any time with little warning, thus reducing the reliability of the work load in the design offices as an indication of future demand.

The industry can do little to stimulate demand by advertisement or introducing 'bargain offers', and in general its only recourse is to lobby the parliamentary policy-makers in the hope of encouraging them to release more public funds and to offer inducements for investment in the private sector. Although other manufacturing industries are faced with similar problems, they normally have greater opportunities for conducting market research into the acceptability of a new product and are less tied to the public purse. The other direction open to the construction industry is to seek work in other markets abroad, and this has been the only alternative path open to the United Kingdom construction industry during the mid 1970's, which have been years of economic cut back at home. In this instance the demand is rather for the production expertise than for the product itself, which may emerge from an entirely alien source.

In the normal situation high demand is reflected in high prices, and an easing in the availability of mortgages, for example, will result in an upsurge of activity in house purchase and a consequent rise in prices. This will further be influenced by the available stock of unsold dwellings and land ready for development. Indeed, the latter could well prove the greater restriction in meeting an increased demand, as the processes of locating, purchasing and obtaining permission to develop land are usually much slower than the erection of the dwellings themselves.

Commercial and industrial development is closely related to the confidence of potential clients which encourages them to accept innovation and change, with its accompanying risks. A buoyant economy governs the relative or absolute growth in demand for buildings in this sector, whilst investment is further influenced by taxation policies on returns, whether in the form of rents or increased industrial efficiency.

One strength of the construction industry is its heterogeneous nature, which enables it to undertake all classes of work through the special mix of skills which it embraces. Its resilience lies in its ability to be flexible, to transfer its fields of work from one product to another with comparatively little manoevring, though inevitably with some loss of efficiency and wastage of resources and skills.

It should be readily apparent from the foregoing paragraphs that assessing demand is peculiarly difficult in an industry as closely dependent upon national financial policies as the construction industry, in all its parts from design office to material manufacturer. This does not mean that within sensible limits it is impossible to make certain predictions for the continuity of employment of existing resources. However, the level of uncertainty is high and cannot be much reduced by the application of the most reliable market research techniques as the customer can usually be identified but those factors which are likely to stimulate his needs are generally unpredictable.

The uncertainty from which the industry suffers and its attendant problems of investment in resources for meeting demand cannot be better demonstrated than in the case of the sudden reversal of policies on industrialised building methods following, though not necessarily entirely consequent upon, the Ronan Point explosion. Encouraged by market indications considerable sums of money had been invested in equipping large factories for the manufacture of prefabricated concrete units on a large scale and in the purchase of sophisticated plant for site casting and erection. In the event these factories were rarely utilised to 60% of their capacity, and probably the most economically viable ones were those installed on sites to meet the sole requirements of those sites, thereby losing the advantages to be gained from a production line servicing a number of outlets and therefore ensuring a reasonable continuity of demand.

Part of the demise of large prefabricated unit manufacture must be blamed upon the inherent lack of flexibility in the design of so many of the products, which could only adapt to low rise dwellings or other building types with the utmost difficulty. Nevertheless, the general concept was never founded on the assured level of demand which other industries would anticipate before investing in such definitive production undertakings, as there was only evidence of government encouragement on a short-term basis and little sign of a guaranteed continuity of demand. The industry is thus naturally reticent in its willingness to invest large amounts of money in its production equipment, other than in the manufacture of building materials. The practice of hiring plant is a characteristic method of meeting fluctuating and changing markets and is not often found in other industries. The pattern of work also encourages the employment of sub-contractors, thus avoiding the use of a large permanent labour force with its problems of continuity of work load, or constant recruitment on a large scale.

QUALITY CONTROL

The nature of construction work makes the introduction of quality control on a quantitative basis more difficult than in the case of most manufacturing industries. The definition of quality depends on:

(a) the properties of the product itself

(b) how well it fulfils the requirements of its use.

If it is to be controlled, quality must be measurable, using such criteria as:

(a) durability or reliability

(b) minimum maintenance cost

(c) precision of dimensions

(d) structural stability and strength

(e) environmental performance (eg, insulation, etc).

These are generally applicable to construction but the continued presence of trade skills and the use of partly processed materials means that in many cases the inspector must rely upon his own subjective judgement and his personal interpretation of specifications. Criteria such as "acceptable appearance" still exist. Quality must relate to design decisions and the designer should be clear in his own mind of the requirements which need to be met in use.

If there is to be control of quality there must be a standard by which to judge a product. There must be a common factor of reference if two products are to be compared for quality, though this reference standard need not be universally applicable. The definition of standards is essentially a communication process with its attendant problems of adequate qualification and interpretation between all those groups concerned in the process. Inspection can be effective only if the specification is unambiguous and is capable of being achieved by the production and assembly process envisaged. Thus the definition of quality is a prerequisite of both its achievement and its measurement, and therefore of its control. Figure 2.3 illustrates the place of quality control inspection in the production cycle.

Figure 2.3   **Location of quality control in the production cycle**

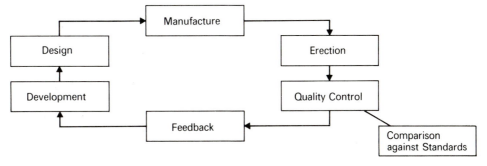

Quality control is the charting of information derived through the inspection of a production process in such a way as to draw attention to the quality of the products while they are actually being produced as distinct from testing completed products and rejecting those below standard. Unfortunately

35

the multiplicity of operations which make up a construction process cannot all be inspected whilst they are being performed and their nature does not usually permit a simple "go/no go" system of testing as can be applied to a set of door furniture or an electric light switch, or indeed to ensure that the cold water tap provides cold not hot water. This means that when completed work is found to be below standard, it is necessary to take it down and repeat the operation. It is clearly essential that any major item of work should be inspected in the early stages to set a standard which will then act as a reference for the execution of later work. The sample panel of brick-work is a typical case in point; but how many in a bricklaying gang even know that it exists, let alone where it is situated?

If it is to be effective quality control must be a planned, continuous process of verification. The system should define responsibilities at all levels. A system must cover all sections of work. It is as much the duty, and need, of the contractor to check work as it is of the designer's representatives. Quality control and inspection should be carried out on a basis of co-operation between the contractors and the designers. With increasing use of specialised techniques and sub-contracted operations this becomes more important. Table 2.1 illustrates the likely benefits to be derived.

The effects of lack of adequate control may be summarised as follows:

1. Failure to perform adequately, resulting in decreased user satisfaction or increased maintenance costs.

2. Structural failure in extreme cases.

3. Loss of production for the contractors due to extensive rectification operations.

4. Loss of production for the contractor due to loss of production flow.

5. Slower production due to lack of feedback of information concerning problems relating to errors occurring in prefabrication, involving additional assembly work.

When the nature and occurrence of defects is examined, a common pattern can be observed between construction and other industries. They generally fall into the following four categories:

1. Incorrect design and construction technology.

2. Defects in materials — inherent weaknesses, faulty preparation, wrong application, poor workmanship, bad storage.

3. Dimensional inaccuracy in setting out, in prefabricated components and in site operations. Instrumentation requires careful consideration in, eg, inherent inaccuracies in the use of tapes.

4. Inadequate protection of completed work and stored materials.

Advocates of the introduction of more rigorous quality control systems have much difficulty in persuading top management of its desirability as there are many problems to be encountered in equating the cost of maintaining an adequate quality control system with the potential saving to the contractor. Manufacturing industry's use of statistical sampling techniques to form an economical basis for inspection can scarcely be used in construction, except in the often quoted case of concrete testing. Few records are kept of what low standards cost to all parties: remedial work is rarely allocated to detailed causes. However, improved communication of quality requirements between all parties costs little.

Table 2.1   **Benefits offered by co-operative quality control**

For the contractor

Effect of control	*Benefit*
1. Fewer defects by reducing the incorporation of poor materials and workmanship.	Reduced costs during progress. Less 'maintenance' costs.
2. Early appreciation of future problems.	Better progress resulting in reduced cost.
3. Better planning.	Integration. Avoidance of delays.
4. Acquisition of data for the solution of future problems.	Confidence. Better supervision.
5. Liaison with consultants.	Communication and mutual appreciation.
6. Enhancement of reputation for supplying satisfaction.	Publicity. Attraction of further work (negotiated).

For the designer and the consultant

Effect of control	*Benefit*
1. Progressive assurance that requirements are being supplied.	Confidence. Client satisfaction.
2. Enables problems to be solved quickly and decisively.	Better efficiency in supervision. Faster progress.
3. Acquisition of practical data about the adequacy of design.	Improved design of future projects.
4. Permits a progressive record of the work especially the causes and nature of omissions and deviations.	Positive direction. Comprehensive knowledge of the work. Limitation of increased costs.
5. Closer association with contractors. Exchange of ideas.	Improved communication. Greater appreciation of site problems.
6. Elimination of contractor's resentment and contempt of 'watch-dog' inspection.	Respect. Co-operation.

At each stage of the process of a building project, it is essential that the requirements for quality awareness and control are met. These reflect responsibilities. Examples at each stage are as follows:

The client: overall responsibility of the safety and welfare (in the broadest meaning of the act) for those who use his building.

The designer: awareness of performance requirements, philosophy of realistic specification and checking procedures.

The manufacturer: sharp quality consciousness; processes and skills related to quality levels.

The constructor: understanding of quality standards specified and the processes of their achievement; awareness of the skills needed for production and its supervision; involvement of the operation.

The supervisor: identification of critical inspection gates; importance of communication channels; recording system for realistic feedback.

BIBLIOGRAPHY

1 STARR, M K. Production management systems and synthesis. Prentice-Hall, 1972.
2 TOWILL, D R and COLLIER, P I. Simple dynamic models of engineering production systems. Work Study and Management Services, pp 388-395, June 1973.
3 ETTINGER, Van J and SITTIG, J. More through quality. Bauwcentrum, 1965.

3. The Contracts Department

The contracts department is responsible for manufacturing the firm's product or more usually products, since many firms have more than one production section. The department has to be geared to produce a pre-determined volume of work since the balance between the finance available and the actual production must be in sympathy. Too little work being done means idle cash, too much work means inability to pay accounts. A company can operate for a time without making profit but it cannot operate without cash.

Even if the demand for contractors' services is strong, it would be irresponsible for the marketing section of a company to obtain more work than the contracts department can carry out, with the financial resources available. Similarly, the 'service' departments such as estimating, purchasing, accounts, etc, are set up with staff and resources to cope with a pre-determined level of production. The inflation of recent years has made it necessary to increase the value of work done in order to remain in the same relative trading position. The liquidity position also needs adjustment to pay the increased wages and increased cost of materials. If inflation continues, there is no easing of this problem.

Organisation of a Contracts Department

There is no common pattern for the organisation of a contracts department. Indeed, the functions which are carried out can vary considerably. The contracts manager in the small works department of a large firm may have a very different role to play than his counterpart in the 'large contracts' section. The client using the small works department may specify the contracts manager he wants when negotiating the work. Agreements will be reached by the client and contracts manager as the work is progressing. On larger jobs, making changes is a much more formalised matter. Indeed, in some of the very large contracting organisations there is no contracts department.

Site problems related to materials are then referred to a materials control section. Plant requirements are dealt with by the plant department and labour organisation is co-ordinated by a labour control section. The services of sub-contractors are arranged by the buying department and the assessment of how well the site is performing is made by an area manager.

Also on large contracts, the contracts manager is at the hub of a very intricate communication system. He is concerned with receiving and transmitting information to:

(a) the design team

(b) suppliers and sub-contractors

(c) site management

(d) the service departments of his own company.

It is not unknown for a contractor to set up a contracts department to deal with the work of a single client. The client might be a manufacturing or a retailing organisation, bank or insurance company. In these circumstances, the client's design team will benefit considerably, when developing the design of a project, from the expertise which the contracts staff have gained from familiarity with that particular type of building.

In low rise housing, the contracts manager will be interested in numbers. His main concern may be the number of stages of construction completed each week. Matching the production to the sales programme is of paramount importance. This involves close liaison with sub-contractors regarding productivity and with clients concerning the provision of sales information on such matters as the colour of bathroom suites, type of floor tiles, etc.

The technology of low rise housing is unlikely to be very time consuming as far as the contracts manager is concerned. His colleague on a large building, may be heavily involved in helping technical specialists such as the site based services co-ordinator, or materials co-ordinator. In some districts, and according to the particular times, a contracts manager may become involved in negotiations with trade unions, perhaps smoothing out or agreeing minor points which are unique to his contract. At the more turbulent end of the scale, he may be working hard in order that his sites do not become involved in an industrial dispute. In many large firms, he would be assisted and guided in these duties by the Industrial Relations Officer.

Organisational form

It can be appreciated that many factors can influence the sort of organisation which a firm develops. The organisational form will influence the duties carried out by the contracts department manager.

Some of the influencing factors are listed in Table 3.1.

Table 3.1   **Influencing factors**

The nature of the product.
The client's expectation of the firm.
The complexity of the technology employed.
The competition.
The size of the firm.
The strength and attitude of organised labour.
The availability and quality of labour.
Changes and trends in modes of employment.
The firm's policy regarding sub-contracting and direct labour employment.

More remote factors
Current legislation.
Foreseeable changes in legislation.
Foreseeable changes in technology.

Table 3.2 shows typical differences in production departments and service departments between small and large firms. In the case of the production departments for the large firm, some of these might be formed into separate entities as subsidiary companies. In this event, they would likely have some, or all, of their own service departments.

Table 3.2   **Typical differences in organisational form**

Departments of a small firm

Production   – New contracts of more than £40 000 value.
                  New contracts of less than £40 000 value.
                  Repairs and maintenance
                  Decorating

Services   – General management – all services

Departments of a large firm

Production   – Contracts
                  Housing
                  Small works
                  Decorating
                  Joinery
                  Electrical
                  Roads and sewers

Services   – Buying
                  Plant
                  Work study
                  Personnel and training
                  Estimating
                  Surveying
                  Production control
                  Accounts
                  Marketing

Figure 3.1 shows an organisational chart for a large firm with a turnover of £30 000 000 per annum.

Figure 3.1   **The Firm**

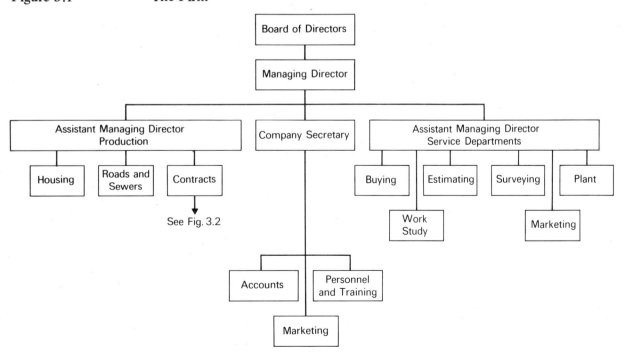

41

Figure 3.2 is the organisational chart for the Contracts Department of the firm. Note that contracts managers 'B', 'C', 'D', 'E', 'F', 'G' and 'H' are each responsible for a number of sites, typically between three and six. This will depend on the geographical location of the sites, the type of work, the architect, and the time sequence in which the work is 'won', ie, one contract manager might find it difficult 'launching' three new contracts simultaneously.

Contract manager 'A' is responsible for one site only. This is a new university to be completed within four years at a cost of £25 000 000. This project's financial contribution to the firm is in the order of £6 000 000 per annum and therefore represents 20% of the current annual work load.

Figure 3.2   **The Contracts Department**

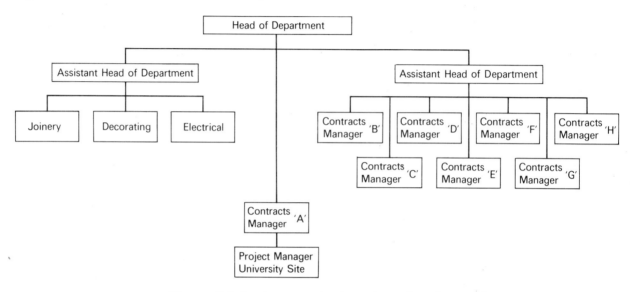

Figure 3.3 is the organisation chart for the production section of the University site staff, and Figure 3.4 shows the organisation chart for the University site administrative and other service staff.

Figure 3.4   **University project (site services)**

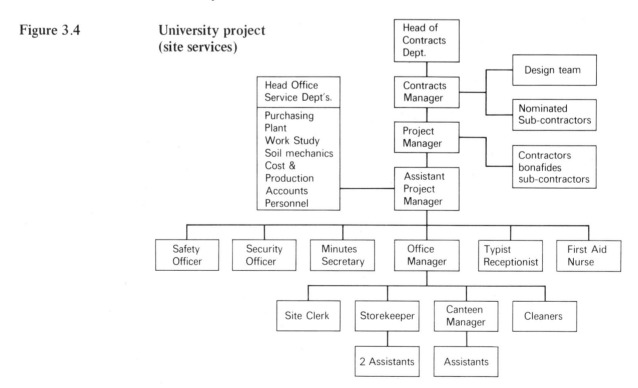

Figure 3.3 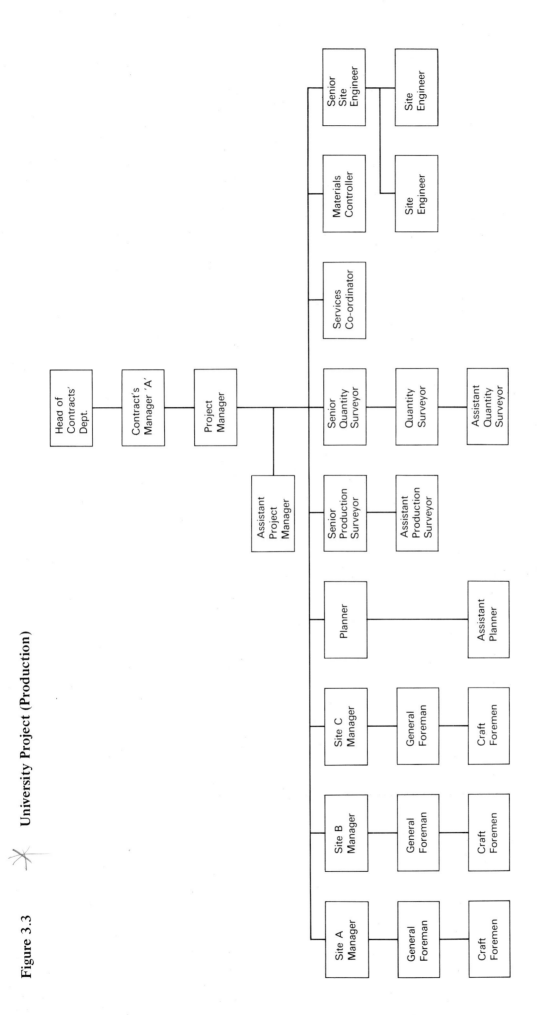 **University Project (Production)**

43

The Manager of the Contracts Department

It is not unusual for the head of the department to be a member of the board of directors and he is frequently given the title of contracts director or production director. This is a contradiction of terms since as a director he is a full member of the board which is charged with the total responsibility of piloting all the affairs of the company. In his capacity as a manager of a department, he works under the control of a managing director as a manager; his job is to implement the policy of the board with regard to the running of his particular department. But, since most customers like to feel that they are getting high level treatment, to have one's affairs handled by a contracts director in preference to a contracts department manager, is perhaps reassuring for them.

The manager's duties

In analysing the duties of the manager of the department in a large firm of contractors, it can be seen that he is linked to three levels.

1. The board of directors.

2. The senior executives' committee.

3. Leader of his group of contracts managers.

As a member of the board, he will help to shape policy regarding markets and products, the forecasting of work load, budgets and profit. He will participate in activities outside the firm to promote its image and to take a share of the social responsibilities of the organisation.

These might include:

1. Involvement in the work of the National Federation of Building Trades Employers.

2. Involvement in the work of the Institute of Building.

3. Interesting himself in the work of universities and colleges, management and craft training.

As a senior executive of the company, he is concerned with his colleagues in developing strategies to match the policies and objectives set by the board of directors. This will include much budgetary planning, co-ordinating the work of all departments. Interdepartmental communication systems are extremely important and are focused around three points:

1. Getting the work.

2. Doing the work.

3. Getting paid for the work.

Control information between the Contracts, Buying, Surveying, Estimating and Accounts departments, is vital to the safe and prosperous operation of the company.

As leader of the Contracts Management team, he will perhaps, wittingly or unwittingly, be influenced by the statement, attributed to the famous French military leader, Marshal Foch, *"A good general does nothing, sees that everyone else does everything, and sees that nothing goes by default."* A conceptual approach which perhaps should apply to the manager of the contracts department more than any other role.

A major aspect of his work will be in employing the right staff around him. This will include the correct number according to the forecasted work load and the best total experience between them. Team building is not accomplished overnight and requires patience, skill and forethought.

Contract manager

The allocation of contracts to contract managers is not always straightforward. If the contracts are allocated to contract managers on a geographical basis, to minimise the amount of distance travelled, then some will manage contracts for which they are not particularly suited. If they are allocated on the basis of the right man for the job, then clearly much travelling will be done and some will have too many jobs to control and some too few. The question is one of striking the best compromise.

Site managers

The head of the contracts department will also take an interest in his site managers. Putting the right man in charge on site according to the needs of the job requires care.

Having the right number of site managers available is something of a lottery. Two contracts, each of £500 000, may require a manager on each site. A single contract of £1 000 000 in value may require one manager on site. In both cases, the contribution to total work load is the same.

At this level, the experience of site managers tends to vary considerably. A man knowledgeable in ground works may have little to offer on a refurbishing contract. In practice, some men do specialise in ground works and transfer to other sites when their section of the work is completed. Some site managers are transferred from one site to another when the level of activity is running down towards the end of the contract. A site manager usually has a special contractual arrangement with his firm; this is often three months' notice on either side for termination of employment.

Some site managers progress to contracts management, for others site management is a terminal role. The role is a demanding one and some find it too tiring when they reach middle age, or perhaps, because of ill health, look for an indoor job. However, due to the high degree of specialisation in large construction firms it is not always easy for the employer to arrange a transfer from site to head office.

Products of their environment

The building site environment tends to breed men who are self-reliant. There are so many variables associated with building, including unpredictable bad weather, sub-contractor failure to perform, labour shortage, lack of design information, and poor design from a buildability point of view, that not surprisingly a site manager often feels frustrated.

Yet the site manager must carry on and build. If he appears sometimes to lack trust in others, it is not surprising. One of the less favourable tendencies of this lonely role is that he may veer towards becoming dogmatic and inflexible and might not take advantage of opportunity to retrieve a contract that is going "off course".

Salesmen

If the client is a visitor to the site, the opinion he has of the site manager is very important. As senior company representative on site, the manager reflects the company and the winning of the next tender may, in the final analysis, depend upon what the site manager says to his visitor.

Leader

The general foreman and the craft foreman, which also include foremen of directly employed men and sub-contractor foremen, look to the site manager for leadership. The weekly site meeting is often the best opportunity for co-ordinating their efforts and improving motivation. Meetings need planning, even though the plans may never be put down on paper. Such a meeting is an opportunity to look backwards at the failures and successes of the past week and then look forward to the production for the coming week. A well motivated site manager with skills in leadership will soon inspire involvement and commitment from his foremen.

Administrator

Many of the administrative functions will be done on a daily or weekly routine basis. Dealing with correspondence is an example of a daily duty. In dealing with suppliers' correspondence, most large firms have standard reply forms to 'standard' situations. The form can thus be completed quickly, although collecting the required information may take a long time. The preparation of time sheets for wages must be co-ordinated with head office concerning when they are completed and delivered so that wages can be paid on the agreed pay day.

Quality controller

Although craft foremen have a responsibility to see that the work done by their gangs is properly carried out and the site engineer to make sure that the building is dimensionally correct, the site manager still bears the ultimate responsibility. To discharge this, he must be fully conversant with the specification and the drawings and have discussed and agreed quality standards with the respective foremen prior to commencing any task. It follows that he must be acquainted with all crafts and learn as much as possible about assessing quality standards.

Negotiator

Recent legislation has clarified the position of the trade unions. Site managers of the future are much more likely to negotiate with the shop stewards and job steward than in the past. For many, this will mean further training in the art of negotiating and is an area of operation in which the site manager will recognise three levels of decision making:

1. When to act.

2. When to act and report the action taken to head office.

3. When to report the business to head office and await instructions.

The contract surveyor

The contract surveyor is concerned with the financial control of building work. Depending on the size and rate of activity of work, he may be resident on site or based at head office and controlling the financial outcome of two or three sites. In most cases, he will be expected to work alone. Since he is dealing with figures, a surveyor must be methodical. Much of his work will be in accordance with The Standard Method of Measurement and the Joint Contracts Tribunal Form of Contract. It is essential that he is fully familiar with the former and with the relevant parts of the latter.

Information collector

One of his major functions will be collecting information for submission to the professional quantity surveyor with regard to his preparation of the monthly interim certificates of payment. In a perfect world, each instruction from the architect would be measured immediately by the professional quantity surveyor and agreed with the contract surveyor. Because this seldom happens and memory is a treacherous companion, a good surveyor keeps records. These records may be supplemented by photographs, preferably signed on the back by the clerk of works and dated. When the surveyor can keep his financial records up to date, it facilitates considerably the accurate assessment of profit or loss, cash flow forecasting and the preparation of the final account.

Dates	Dates assume tremendous importance in contract work. This particularly applies to keeping records of all delays caused by the late receipt of information, the late delivery of goods by nominated suppliers and other incidents outside the company's control. He would report to his superior any difficulties arising in connection with the agreement of claims and interim valuations.
	A regular routine is the weekly or monthly, according to agreement, measurement of labour only and appropriate domestic sub-contractors. The monthly check-up will normally include a cost and value comparison.
Negotiator	When working with the professional surveyor, his role will be that of negotiator on behalf of the contractor. Good diction and the ability to communicate are essential prerequisites for this role.
Self discipline	Being a surveyor is a position of trust. It is not easy for anyone to check on the accuracy of his work. Although the JCT Form of Contract refers to the measuring of all variations by the professional quantity surveyor there tends to be differences between theory and practice. It is considered, by some people, that the professional surveyor is under no obligation to draw to the attention of the contract surveyor deficiencies in his claims.

General foreman

The traditional route to this position used to be through craft training to craftsmanship, then to craft foreman and on to general foremanship. Many firms now prefer site managers and contract managers to have a period as assistant general foreman and then general foreman on their way up the career ladder. It is becoming increasingly common to find general foremen who have not had craft training but taken up the post after more general building studies.

Safety	It is his responsibility to ensure that all methods of work are safe, and that no member of the public or the work force is put at risk. This includes supervising the maintenance of hoardings, gates, site boards and pavement cross-overs in a safe, clean condition.
Programming	It is his duty to be responsible for short term programming. This requires arranging and supervising the 'taking off' of operational quantities or dealing with this personally.
Materials	The scheduling and calling forward of materials and plant to match the programme requirements.
Labour controller	Of the management team, he is in the best position to judge the suitability or otherwise of the labour employed. When operations are drawing to a close, he will recommend the transfer arrangements to the site manager. When operations are commencing, he will make recommendations concerning the recruitment, selection and engagement of labour.
Communicator	The site manager is in constant touch with the general foreman who communicates the site manager's instructions to the various craft foremen.
Recorder	All general foremen should maintain a site diary for record purposes. Where a proper costing system is employed, the general foreman will check daily that the labour allocation sheets completed by the craft foremen are an accurate record of the work done that day and that hours worked are correctly referenced to the cost coding system.

Special qualities   The building labour force is often a diverse group of individuals and the special quality which one hopes to find in a general foreman is the ability to motivate them all to get on with the job. To achieve this requires an outward looking person with an optimistic and positive approach to work. His instructions to craft foremen will usually be given verbally, so pleasantness of manner is a useful attribute.

Further job descriptions of contractors' staff are given in the Appendix at the end of this book.

4. Pre-Tender Planning

Assuming that the contractor has been invited to tender for a project under the JCT Form of Contract, the first point which must be established is the 'relative worth' of the contract to the company.

This cannot be assessed by intuitive means and a carefully planned investigation should be made. The procedure and sequence of such an investigation should be varied according to the particular needs of each individual contract. The outline which follows is intended to portray typical arrangements and is not necessarily suitable for all circumstances. It is quite likely that an estimator will be asked to extract from the contract documents a brief statement of facts.

Contract appreciation – Stage 1

1. Name, business, address of client.

2. Name and address of the architect and any other named consultants.

3. A brief description of the project which would include:
   (a) the site address.
   (b) brief particulars of the construction including
       (i) the constructional system
       (ii) the area of the building(s) and number of floors
       (iii) the site plan.

4. The contract dates, commencement, completion, defects liability period, liquidated and ascertained damages, conditions related to payments.

As soon as possible after these facts become available, a preliminary meeting with some of the senior executives should be arranged. Typically, this might include the managing director, the head of the contracts department, and the company secretary. Once they have examined and discussed the project it may be clear to them that it is unsuitable for the company.

In this event, they will decline the offer to tender, return the documents and that will be the end of the matter. However, if the job looks attractive, they will decide what further information they require, the time scale for its preparation, before finally agreeing whether to tender or not.

Contract appreciation – Stage 2

The client
Obviously the contractor would want to feel confident that legitimate requests for payment would be honoured, therefore, it is necessary to make enquiries about the client's business record. It is also desirable to ascertain if his approach to working with the architect is likely to be satisfactory. The client would be expected to supply the names of bankers, etc, to whom reference could be made.

The architect and consultants
If any members of the design team are unknown to the contractor, enquiries should be made as to their professional competence and also whether they have previous experience of working together. If they have, it is reasonable to suppose that the selected contractor will have fewer communication problems.

The contract documents
All the documents should be scrutinised with a view to detecting anything unusual. Examples which fall into this category are:

(a) unusual contractual arrangements regarding payments

(b) restrictions on the use of the site

(c) phased handover problems

(d) unusual forms of construction

(e) the degree of repetition of construction

(f) unusual insurances

(g) 'rogue' loads, ie unusually heavy or difficult objects to handle.

Should there be any unusual aspects, the risk must be quantified and the cost assessed thus ensuring that due allowance of time and cost is made if it is decided to submit a tender.

The project
The contract's manager must assess whether his team have the management expertise and other necessary resources to carry out the project. If it is highly complicated relative to the quality of the firm's management expertise, the possible solutions should be analysed. If there is a probability of more of this type of work arising in the future, he may wish to use the project for gaining experience.

Approximate value summary

By examination of the drawings and other documents, the approximate value of the contract can be ascertained by approximate estimating techniques. If the value of prime cost items and nominated sub-contractor's work is deducted from the total, then the value of work to be carried out by the contractor and his sub-contractors can be obtained. If this is considered against the contract commencement and completion dates, it can be seen how this will fit in with the work load pattern (see Table 4.2).

Decision to tender
At this stage it will generally be possible, and necessary for reasons of cost, to decide whether a serious tender should be submitted. Apart from the factors already mentioned, the following require consideration.

1. The value in orders and work in progress the company already has on its books and how desirable the project under consideration is.

2. The probable competitors and how keenly they are likely to compete.

3. The work climate and the future. If the amount of work coming on the market is declining, it might be advisable to strive for volume of work at low profit and maintain a good estimating success ratio.

4. It is essential to try to maintain goodwill with architects and clients. This may mean tendering for, obtaining, and carrying out work which is not particularly attractive to the company.

Table 4.1   **A pre-tender procedure check list**

Item	Person responsible for action	Date required
1. Check client architect quantity surveyor approximate value competition liquid and ascertained damages.	J Smith, estimator	x
2. Check drawings, inaccuracies) complexity of work) degree of repetition) major construction form) at architect's office	R Jones, planner T Fox, contract manager	x x
3. Prepare site investigation report	T Howard, estimator	x
4. Prepare methods statement	(T Fox, contract manager (R Jones, planner (T Howard, estimator (L Smith, plant dept (W Snowdon, buyer	x x x x x
5. Prepare programme	R Jones, planner T Howard, estimator	x x
6. Prepare site layout schemes	R Jones, planner	x
7. Check labour availability in the locality of the site	Personnel officer	x
8. Determine work to be sub-contracted	Head of contracts department T Fox, contract manager	x x
9. Site organisation structure	T Fox, contract manager	x
10. Check materials supply position, prices and quality Negotiate with sub-contractors, determine fixed price allowances	(W Snowdon, buyer (((	x
11. Contract conditions Contract insurances	(A Fanning, company secretary (	x
12. Pricing the bill* (This activity will be continuous from the decision to tender date)	T Howard, estimator	x
13. Preparation of the estimate	T Howard, estimator T Fox, contract manager	x x
14. Preparation of the tender	Managing director	x

*The making of this decision would be preceded by a meeting of the senior personnel involved in the contract.

Pricing the bill

After examining the documents the estimator will have a shrewd idea of the likely performance figures of craftsmen engaged in bricklaying, plastering and many other craft operations. However, he would be unable to price these items until he knew by what means the materials were to be handled, ie bricks delivered palleted and handled by fork lift truck will not necessarily cost the same amount of money as bricks barrowed and delivered by platform hoist to the bricklayer. Long runs of continuous work with a high degree of repetition and consistency in the methods used all help to keep costs down. The estimator will need to know how the contracts department envisage work being done before it can be satisfactorily priced.

Some of these matters can be resolved after a few minutes' discussion between estimator and planner, others will require very detailed consideration by the planner before the method of approach to work is finalised. For example, it may seem obvious, at first glance, that the concrete for a suspended cast in-situ floor should be pumped, but if a tower crane is essential for hoisting and erecting cladding panels, the calculation is not quite so simple. The following factors may have to be taken into consideration:

(a) the extra cost of the pump

(b) the saving in preliminaries due to a shorter contract period

(c) the increased cost in formwork for the suspended floor if pumped

(d) the availability of plant and equipment

(e) past experience of the site staff in this type of work.

It will be appreciated that a normal bill of quantities is not a very useful document to a planner. The planner prefers to have his quantities in the operational form by which the work will be done. For example, assuming that the intention is to dig a large basement excavation by drag line and to revert the sides of the hole to avoid using timbering, the amount of earth dug will bear no relationship to the quantities in the bill. The price of each cubic metre mentioned in the bill must bear its fair proportion of the cost of the extra digging.

Planning, estimating and buying

Whilst it is not the intention of this book to dwell upon the specialist work of estimating, the close relationship of the estimator's work and that of the production planner cannot be too highly stressed. The estimator needs to know how the planner anticipates the work being done. There is also a strong interrelationship between the work of the estimator and that of the buyer. In the larger company, the estimator will rely very heavily on the services of the buyer. In the small company, one person may carry out the three functions of planner, estimator and buyer. Quotations are required from not only suppliers but also sub-contractors. In periods of great instability, almost all prices should be checked.

A skilled buyer should be in a much better position than an estimator to know who will quote competitively. He will also know which sub-contractors and suppliers will quote competitively in certain areas and for particular types of work.

It is normally the responsibility of the contract manager to decide which trades will be sub contracted. It is fairly usual practice to separate the various trade bills and to make photocopies of the relevant pages of those bills for which a price is required. The number of sub-contractors invited to tender will vary with the size and nature of the contract. It is customary to send the sub-contractor two copies of the sectional trade bill, one for his retention and one to return to the contractor. It will also be necessary to supply each potential sub-contractor with a brief 'description of contract' statement. This will state whether the contract is fixed or fluctuating price, the date for the return of the quotation, the approximate period when the work will require to be done. Similar arrangements are carried out with suppliers but only one copy of the relevant section of the bill is sent since suppliers will normally quote on their own standard forms.

**Interaction —
methods statement,
pricing the bill,
site layout,
tender programme**

The nature of the interaction between the work of the estimator, buyer, and planner has already been stressed, but there is an equally strong affinity between pricing the bill, the preparation of a methods statement, the site layout scheme, and the tender programme. Generally speaking, the preparation of the methods statement precedes the other three activities, although it is noteworthy that there must be enough space to adopt the method (site layout) and enough time (programme) to carry it out.

At a typical pre-tender planning session, the contracts manager would describe to the planner and estimator how he envisaged the work would be tackled, and the order in which operations would be commenced and completed.

The methods statement

The methods statement is the statement of intent of the contract manager's planned way of working with regard to the major items of construction. This could include parts of construction which are relatively small in time content but require forethought because they are complicated.

The amount of effort and enquiry given to the preparation of the method statement will be dependent upon the following factors:

1. The location and nature of the particular site and its immediate environment.

2. The amount of site space to establish storage and work place areas and deploy plant.

3. The company's expertise in that type of construction.

4. The availability of resources.

5. Company objectives such as the least possible construction time or maximising on the use of certain resources which will be available for limited periods only.

6. The complexity of the site, ie access, shape, slope, storage areas available, type of soil, or unusual geographical or environmental factors.

7. The opportunity available for technique development.

Typically, a meeting would be arranged between the contracts manager, planner and estimator. The contracts manager would outline such matters as:

(a) the commencing priorities for the contract

(b) at which part of the site he would commence operations

(c) the plant and other resources required to carry out the contract in the way he envisaged

(d) the site accommodation and workshop areas.

The planner would write down at the meeting in a suitable form (ie circle and link diagram, or network diagram) the contract manager's intentions. After the meeting, he would check the validity of the contract manager's intentions. The following points would receive attention:

1. The construction logic.

2. The capability of the plant resources to carry out the work.

3. The safety of the proposed methods.

4. Difficulties in providing continuity of work.

5. The possibility of attaining the required quality standards by the proposed methods.

6. Avoiding carrying out work at unseasonable times, ie such as external painting in January.

Availability of the resources

The estimator would provide information from company records on the likely output figures to be expected. If alternative methods have been discussed, he would carry out a study of the two methods to determine the probable cost and time factor for each method.

Site layout

The next duty of the planner would be to prepare the pre-tender programme and supporting schemes for site layout(s). The preparation of these would proceed simultaneously, but for reasons of clarity to the reader, they are presented separately.

The site space is used to create a temporary workshop from which the contractor will erect the building. There is no standard ratio between the space available for site work places and the space for the proposed building. Sometimes very tall buildings are erected and the contractor may have almost no space to operate from. Contra-wise, he may have just a few small buildings to erect on a large level site. Each site must be considered as a new problem. The mode of employment of labour may also affect the site installations. If the intention is to employ all, or nearly all sub-contract labour, the main contractor may be reluctant to make extensive arrangements which lead to high productivity unless the sub-contractors are prepared to recognise this expenditure by reduction of their rates. In most instances, they are unlikely to agree to this.

To achieve the optimum profit from a project, there must be a correct amount of money expended on site installation. Too little expenditure results in unnecessary double handling, excessive walking about, and many other wasteful practices which lead to low productivity. Too much planned expenditure leads to either not winning the contract or winning the award and making inadequate profit.

Site layouts are developed around five facets:

1. The movement of men, plant, components and materials.

2. Accommodation for personnel, both welfare and working accommodation, the storage areas and buildings for materials.

3. Work activity such as fabricating steelwork or formwork, mixing concrete, etc.

4. Site access for the passage of vehicles of varying characteristics.

5. The control of work activity.

All of these aspects have cost implications which the planner should study carefully and inform the estimator of his findings so that proper allowances are made in the estimate.

It is in the contractor's interests that vehicles delivering goods to the site can do so without difficulty or delay. If it is planned to use ready mixed concrete, then the site road capability must be weight of delivery vehicle plus weight of load of concrete (typically 6 + 12 tonnes).

If structural steel is to be delivered in 40 tonne loads, the length of the load and the turning circle of the delivery vehicle should be ascertained. This length and turning circle should be reflected in any changes of direction in a site road.

If heavy vehicles will be operating around the site during the construction period and the permanent roads are of light construction, it is usual to use the hardcore base as a surface for the construction plant. This hardcore bed tends to suffer damage and requires re-forming before the road is finished. The estimator must remember to allow for this item of maintenance. If the specification calls for a 150 mm site scrape followed by a hardcore bed of 150 mm, it is advisable to scrape less than 150 mm, possibly only 100 mm. The site traffic will 'punch' the hardcore into the ground, the exact amount is difficult to determine, but is unlikely to be less than 50 mm. If it is intended to put down a site road of lean-mix concrete which is solely for construction purposes, the estimator must allow for its breaking up and removal on completion of the contract.

As the planner proceeds with preparing the pre-tender programme, he will have to consider, as the various stages of construction are reached, how relevant the site layout is to the next stage of construction. He must inform the estimator of any changes he foresees if additional expense is involved in the change.

If men are being transported into a multi-storey building by passenger hoist, careful consideration must be given to the number of hoists provided. Obviously the installation and running cost of a hoist is an expensive item. Projects have been carried out in London with peak labour force in excess of 2000. Assuming the hoist has a capacity of 10, it will therefore require 200 journeys to get the labour force to work. The problem can be eased by adopting staggered starts and finishings but the cost implications of having one, two or three hoists would be considered. Associated problems would be the cost of sanitary provisions and meal and tea break organisation and costs. On some sites, the planner and estimator may have to allow for the time and cost of moving the administrative centre.

Ultimate efficiency depends upon:

(a) minimising double handling

(b) adequate storekeeping arrangements

(c) minimising walking distances

(d) avoiding loss by the elements

(e) adequate security arrangements to combat theft and vandalism

(f) minimising congestion

(g) continuity of employment

Pre-tender programme

The pre-tender programme is a most important document and is increasingly regarded as such by architects, clients, and bank managers. If a contractor is seen by the architect to have prepared this programme carefully, he can adjudge whether the contractor's staff have understood his design. It is important therefore that any person liaising with the architect or other consultants during the pre-tender stage should be fully briefed on the contractor's policy and plan towards the project; he should be a responsible member of the contractor's team and likely to be in the contractor's employ for the duration of the proposed contract.

Whilst all the foregoing activities have been in progress, viz: site investigation, examination of the contract documents, method statement and site layout schemes, the contract management team will have been preparing query lists which will be submitted to the architect for his examination and in some instances to pass the queries on to other consultants.

Points for clarification

These may include:

1. Discrepancies in the description of the same item between the drawings, Bill of Quantities and specifications.

2. Conflicting dimensions or lack of dimension.

3. Unrealistic tolerances or tolerances in conflict.

4. Specification description which is too vague.

5. The contractor's representatives may be aware of serious delivery difficulties of some of the materials specified.

6. Materials may have been specified which are unknown to the contractor's team and they may wish to see specimens or take photocopies of the performance specifications supplied by the manufacturer.

7. Since the commencement of the negotiations, the architect will probably have agreed further nominations and perhaps agreed dates with specialist suppliers and nominated sub-contractors.

Typical of the questions which the contractor's team will be asking at this stage are:

1. Have all the planning approvals been obtained from the planning authority?

2. Have the electricity board (or other statutory body) obtained the necessary way leave to bring their supply across adjacent premises?

3. Has the party-wall agreement been signed?

4. Has the soil mechanics specialist completed the bore hole log?

They may wish to suggest to the architect certain changes of design or detail which would make the project easier from the contractor's viewpoint and in some cases cheaper to the client.

The contractor's team will try to assess whether the architect is proceeding with his work in a businesslike fashion. They will have some indication of this from the standard of communication and working relationship which he has built up with other members of the design team. They will encourage him to provide in writing details of any dates which he has agreed with nominated sub-contractors and suppliers, copies of any contractual conditions which he has entered into, and key dates which he has mentioned regarding the project such as:

(a) date for possession of site

(b) date for delivery of materials in scarce supply which the architect has made the subject of special order

(c) dates agreed with other members of the design team for the supply of drawings and other particulars

(d) phased handover of parts of the construction to the client

(e) any special conditions imposed by the police, local authority, or client, which they have mentioned in discussion with him

(f) any special points regarding the submission of the tender

The planner, supplied with a site investigation report, a method statement, and a schedule of planned performance rates, fully documented by the design team, should be able to go ahead and produce a realistic pre-tender programme. Unhappily, much of what has been described is, in the experience of many planners and estimators, a counsel of perfection in an imperfect world. All too often they are very busy planning and estimating with too little information, too little time and too much competition.

As he prepares the programme, the planner will seek to identify the key operations, and the dependency between operations, those operations which are non-critical and which by their flexibility of timing will enable the contractor to obtain a high utilisation from his resources. He will consider those operations which are critical and must be controlled with the maximum of care. He will look for the 'lead' and 'lag' between succeeding operations, and also for operations which are likely to proceed without interruption, operations which are intermittent and to what extent they are intermittent.

Materials — influence on planning

For reasons which are explained in a later chapter, the building materials industry is seldom in a position to satisfy completely the needs of the construction industry, hence at any given time some materials are in short supply. In certain instances, the architect will have placed orders on behalf of the client before the contractor is selected. Generally, it is the responsibility of the contractor's buyer to be aware of the prevailing market conditions and to highlight where difficulties could occur when he reviews the bill of quantities and the specification.

These scarce materials can have a major influence on a programme. There is little point in preparing the base of a steel framed building in April if steel for the frame cannot be delivered until October. In such circumstances, the contractor would be denied the use of the retention money for an unnecessarily long period. The planner must therefore consider, before establishing the start date for any operation, whether the materials can be obtained before that date, allowing time as required for preparation of the material and site handling.

Site organisation structure

By inspecting the pre-tender programme a schedule can be prepared indicating the site staff required to supervise and control the project. The schedule would show the dates when each person was required on site and when he would be available for transfer to the next. If it appears premature to attend to this sort of detail at so early a stage, it should be remembered that this information has a bearing on many aspects of cost, including:

1. The calculation of the cost of supervision.

2. The calculation of the cost of site accommodation.

3. The greater the number of supervisors the higher the productivity of the operatives.

If an over-generous allowance is made the contract may be lost, yet to arrive at the right amount of supervision is not easy; the following factors have an influence:

1. The number of directly employed site workers.

2. The complexity of the work.

3. The amount of service to be provided by the head office.

4. The quality of the communication system between the design team and the constructors.

5. The speed of erection.

If both site manager and site agent are engaged on the same site, it may be practical and possible to remove the site manager to another site after the main constructional work is completed. This would obviously reduce site overheads. Naturally, this would be done only where one man could reasonably handle the services and finishes.

Finally, it is noteworthy that one of the most difficult aspects of contracts management is having the right number of staff in the department in the right disciplines in order to keep all staff fully engaged.

The commercial decision

When all the pre-tender activities have been drawn together and the estimator has totalled the estimate of real costs, which includes allowances for the use of services provided by head office, it is the responsibility of senior management to decide at what level to set the profit (the mark-up). This is probably the most important decision which has to be made in the history of any contract. The following factors have to be considered:

1. Return on the capital employed on the contract.

2. Risk profit proportionate to the degree of risk envisaged.

3. The environment, and the competition.

Presentation of the tender

If a number of contractors are invited to tender for a project, it can be assumed that the architect believes that any one of them has the capability to do the job. On what basis will the selection be made? Obviously, the lowest price must be a consideration but the lowest tender does not necessarily result in the lowest ultimate cost to the client. The architect will consider the tenderer's 'claims' history. Perhaps more important, he will be influenced by the amount of preparation that the contractor has done and how well he understands the design concept and the client's needs. It is advisable to prepare a document for submission with the tender to demonstrate to the architect and client that the contractor's team fully understand the project. The document should include:

(a) proposed sequence of construction

(b) pre-tender programme

Table 4.2   Work load chart

Contract No.	Contract Sum	Commencement	Completion	Value Completed	Value Uncompleted	Revised Completion Date
1	1,200,000	Nov 1976	April 1980	1,000,000	300,000*	
2	1,500,000	Jan 1978	April 1980	1,100,000	400,000	
3	1,700,000	Jan 1978	May 1980	1,000,000	700,000	
4	2,500,000	Feb 1978	July 1980	2,000,000	500,000	
5	1,000,000	Feb 1978	August 1980	500,000	500,000	
6	25,000,000	Sept 1978	April 1982	12,500,000	14,000,000*	Jan. 1982
7	4,000,000	Nov 1978	Mar 1980	3,500,000	750,000	
8	1,000,000	Jan 1979	Mar 1980	750,000	250,000	
9	2,400,000	Jan 1979	August 1980	1,000,000	1,400,000	June 1980
10	1,700,000	Mar 1979	May 1980	1,200,000	800,000*	May 1980
11	1,300,000	Mar 1979	Sept 1980	800,000	500,000	
12	800,000	Apr 1979	April 1980	600,000	200,000	
13	2,100,000	May 1979	June 1981	900,000	1,400,000*	
14	1,100,000	July 1979	Nov 1980	500,000	600,000	
15	2,400,000	Nov 1979	August 1981	600,000	1,800,000	
16	1,600,000	Dec 1979	June 1981	50,000	1,100,000	
		WORK IN HAND			25,200,000	

(c) section hand-over dates

(d) finance statement showing anticipated payment amounts and dates

(e) a schedule of information required

Architect and client

It is the architect's duty to help the client select a contractor who is capable of building the project at a fair price. Occasionally, he will invite tenders from firms of whom he has little or no first hand experience. Whilst he will assume that all contractors employ expert staff who should be able to prepare extremely accurate tenders, he should take a very cautious approach, if, say, two firms, neither of which he has personally had dealings with, have submitted tenders very much below any others.

With the best of efforts, tender documents seldom depict with complete clarity all that is involved in carrying out the work. Tenderers who have had personal experience of the architect will averagely be better placed for gaining, or making assumptions about, the missing information.

It is very important that the architect should assess whether the contractor has interpreted correctly his intentions; in order to do this he must examine each submission carefully. The more comprehensive the submission, the better he is able to make his assessment. The contractor supplying a programme, including estimated labour strengths, site management organisation chart, a plant summary, a site layout scheme including hoarding, signboard(s), temporary roads, site buildings and site materials storage areas, is much more likely to influence the architect in his (the contractor's) favour than the firm returning a priced bill only.

The architect will arrange his internal organisation to suit the needs of the contractor's programme. If this programme is carefully prepared and adhered to, the architect can assess when the information concerning the various stages of the work will be required and allocate members of his staff accordingly. This is not as easy as it sounds; for example, the contractor may require the landscape drawings before setting up the site in order to identify the existing trees which have to be saved for the ultimate landscape plan.

Summary

A most critical period in the life of any contract is unquestionably the pre-tender period. In almost all instances in this procedure, the contractor is compelled to produce his tender in most unreasonable haste. Often unrealistic rates are given in the bill because of human error, laxness in checking, and guesswork when careful calculations ought to have been made.

In order to minimise these mistakes, most contractors have a 'procedures manual' which provides a basis for carrying out the pre-tender planning. The head of the estimating department will arrange for the allocation of work through the various heads of departments and monitor the progress of the work to an agreed timescale.

Work in hand charts   Table 4.2 gives a summary of the contracts in hand and the value of work uncompleted. The figures for contracts 1, 6 and 10 demonstrate that value completed plus value uncompleted does not necessarily equal the original contract sum for the obvious reasons of 'rise and fall' and additions or subtractions for variations.

Table 4.3   Forecasted quarterly work load from January 1980

Contract No.	Uncompleted	1980				1981				1982				1983			
		1	2	3	4	1	2	3	4	1	2	3	4	1	2	3	4
1	300	200	100														
2	400	250	150														
3	700	400	300														
4	500	250	200	50													
5	500	200	200	100													
6	14000	2000	2000	1800	1800	1600	1600	1600	1600								
7	750	500	250														
8	250	250															
9	1400	700	700														
10	800	600	200														
11	500	250	200	50													
12	200	150	50														
13	1400	300	300	250	250	200	100										
14	600	300	250	50													
15	1800	300	400	300	300	300	200										
16	1100	200	200	150	150	200	200										
	25200	6850	5500	2750	2500	2300	2100	1600	1600								

Table 4.3 shows the planned quarterly completions until 1983 which will see all existing contracts completed. The amount of work in hand and scheduled for completion in 1980 is £17 600 000 which gives a shortfall of £12 400 000 if the firm hopes to achieve a turnover of £30 000 000 per annum.

5. Post-Tender Plann

Architect's project meeting

As soon as possible after the selection of the contractor, the architect will call a project meeting attended by:

Design team representatives
- Architect (Chairman)
- Secretary (member of architect's staff)
- Professional quantity surveyor
- Structural engineer
- Clerk of works
- Resident engineer

Contractor's representatives
- Contracts manager
- Contract surveyor
- Site manager

The Agenda will be concerned with the following matters:

Communications

1. Acceptance or amendment of the programme.

2. Agree procedure for future meetings.

3. Immediate and future information required.

4. Agree procedures for the circulation of information between any parties involved in the contract. The general principle being that all communication should be circulated through the architect since he is the co-ordinator of activity and information flow.

Payments

Agreement will be reached on the dates for the submission of interim certificates to the professional quantity surveyor. This will include the apportionment of items in the Preliminaries Bill, the format and detail required in daywork accounts, and the operation of the formulae price adjustment clause.

Production

1. Discuss and agree which trades the contractor may sub-let.

2. Agree the contractor's provisional site layout. This is particularly important if the client is to retain the use of part of the site as, for example, with a school building already in use.

Planning within the architect's organisation.

The following, extracted from "Management of building contracts" published for NJCC by RIBA Publications Ltd., and reproduced here describes some of the matters to be dealt with at this stage.

1.0 As soon as possible after the appointment of the contractor, the architect, with the contractor and the quantity surveyor, should:

1.1 ascertain the titles and names of the contractor's administrative staff both on and off site, and their respective responsibilities and duties

1.2 ascertain the names of other consultants, including structural and services engineers, and their respective responsibilities and functions

1.3 verify whether or not there will be a resident architect, resident engineer(s) or clerk(s) of works, and their respective responsibilities and duties

1.4 ascertain the contractual duties for possession and completion of the works, and other relevant matters referred to in the contract and contract bills

1.5 ascertain that the contract, contract drawings, and contract bills have been prepared, and are ready for signature by the parties to the contract

1.6 ascertain that necessary drawings, schedules, contract bills, and other essential information are available to the contractor

1.7 ascertain that the architect, quantity surveyor, contractor, and consultants are sending copies of relevant correspondence to each other

1.8 establish the number of copies of drawings, bills of quantities, minutes, reports, etc., that are required by the architect, quantity surveyor, contractor, and, if appointed, resident architect, resident engineer(s), clerk(s) of works, other consultants, subcontractors, and suppliers

1.9 establish details of material to be supplied by the employer, or which works are to be carried out by direct contract

1.10 ascertain that necessary approvals under Building Regulations, Town and Country Planning Acts, and other statutory requirements have been obtained, and that due note has been taken of any special restrictions

1.11 ascertain that quotations have been received by the architect, and checked by the quantity surveyor for items covered by prime cost sums within the contract bills, and that these quotations, in the case of firm price tenders, cover the accepted contract period, and that warranties, if required by the architect, have been given

1.12 establish the dates for future project meetings (chaired by the architect) and the dates for site meetings (chaired by the contractor). (In some instances, the employer may appoint a representative to attend such meetings on his behalf. This will not affect the rights and obligations of the employer, contractor, architect, or quantity surveyor under the terms of the contract.)

Planning within the contractor's organisation

General

Once the contract has been awarded, the managing director must inform all departments within the firm so that necessary actions can be commenced. Arrangements should be made with the architect for the contract to be signed or, in the event of any difficulties, a letter of authority to proceed, to be signed by the client (Employer). Any insurances required under contractual conditions should be obtained.

The Contracts Department

The contracts department will appoint a site manager and involve him as soon as possible in the preparations. The architect should be contacted immediately to find out any further information which may be available. The architect should be supplied with the names and telephone numbers of all company personnel involved with the contract and at the same time details of the architect's and consultant's personnel should be obtained. In the light of further information now available the method decisions can be reconsidered. Approval can now be obtained from the architect for the sub-contracting of certain work. Internal memos should be sent regarding the requisitioning of plant, hutments, safety and welfare equipment.

The site must be viewed to determine the exact position of hoardings, pavement cross-overs, sign boards and necessary applications for these to the local authority. The site layout should be prepared in detail, indicating the position of fixed plant, material storage areas, hutments, including toilets and canteen, temporary roads, pavement cross-overs, hoardings, job sign board. Application must be made to the appropriate statutory bodies for water, electricity, telephone and sewer connections.

The local factory inspectorate, fire authority, police and doctor should be notified of the proposed commencement date of the contract. A schedule of information required can be prepared to include the dates when it is required by. In the light of the foregoing, the pre-tender programme should be developed into a contract programme and arrangements for the transfer of labour and supervisors to the site can be made.

Obviously, these tasks will not be completed unless there is an established procedure within the organisation, placing responsibility for their execution and the sequence in which they should be completed.

The Buying Department

The department will consider the supply of materials, examining availability against the dates required by the programme. Many of the quotations received during the pre-tender stage will be accepted, in some cases they will look for alternative suppliers or substitute materials.

Final check before occupying the site.

Before occupying the site, the contract's manager will make a final check over the following points:

1. That hoarding licence and pavement cross-over approvals have been obtained.

2. Notice of commencement of work sent to H M District Inspector of Factories (Form 10).

3. Prescribed notices, placards and registers are available, including a copy of the firm's safety policy as required by Health and Safety at Work etc. Act 1974.

4. Welfare and safety matters have been attended to, such as the notifications in writing of the persons appointed to be safety supervisor and first aid nurse respectively, Adequate protective clothing, etc. should have been ordered from stores and similarly the provision of first aid boxes, ambulance room equipment, etc.

5. The fire security officer has been selected and has made requisitions for fire fighting equipment, appropriate warning notices, and key board. The officer will have been involved in the planning of fuel storage areas and aspects of the care of combustible materials whilst in storage.

6. If the site is already occupied by the client, liaison will have been established with the client's welfare, safety, fire and security officers and arrangements made for co-operation in the event of an emergency. ncy.

7. Before operations commence it is vital to check that all insurances, and permissions, have been obtained and that adequate record is made of the boundary conditions. These should be supported by photographs where necessary.

8. Services essential to the construction operation, ie, electricity, telephones, water, etc, will have been ordered and a check should be made to ensure that these matters are in progress and will be available on the date asked for.

9. Offices and toilet buildings or trailers will have been ordered from the plant department and a notice sent to the local authority of the intention to erect these temporary buildings. Arrangements will have to be made with the local authority, or privately, for the collection of rubbish, particular thought being given to waste food from the canteen and avoiding rodent problems. In certain circumstances the rating authority will wish to levy rates on the temporary buildings.

   Where twenty or more persons are employed, or where there are ten or more persons employed at any time other than on the ground floor, the Shops, Offices and Railway Premises Act of 1963 stipulates that a fire certificate must be obtained from the Fire Authority to the effect that the premises are provided with reasonable means of escape.

10. In order to have all the above nine points attended to, a carefully scheduled timetable must be prepared. Since there are many dependencies and there is a repetitive aspect, ie most of the above matters will have to be attended to on every site the contractor occupies, it is therefore advantageous to develop and use a standard procedure.

Occupation of the site

Site Security

When all the preliminary actions have been attended to, the contractor will occupy the site. Security is particularly important. In most cases, the first task is to erect hoardings, gates, hutments, lay site roads, electricity and drainage services. Even when it is not the intention of the management to have a permanent night watchman on the site, it is worthy of consideration for this period of the contract only, when access is particularly easy and many unsecured, unfixed goods are lying around. Naturally, the local police should be notified of the date when activity is to commence.

Site office accommodation – types

Mobile accommodation may be extremely suitable on very large area sites, for example, new road construction. A less obvious situation, but also relevant, is during the modernising of housing estates which could extend over many acres.

This choice of accommodation may be the most suitable when it is impossible to find a place where the accommodation can remain for the total duration of the contract. The problem of minimising the movement of temporary services should be given consideration if mobile offices are selected. Office accommodation may sometimes have to be positioned on or under scaffolding. In these circumstances, thought must be given to the safety of the office workers from objects falling on to the site offices or communicating passage ways.

Two storey office buildings naturally reduce the amount of site space occupied whilst at the same time improving the view of 'the works' for the occupants of the upper floor rooms. This can be a very important consideration if the project includes complex ground works.

On most large sites, the contractor is required to provide a conference room; if the client is a government department, it is likely that larger accommodation will be required than a similar project for a private client. In order that the clerk of works has adequate accommodation, the precise size is often stated in the Bill.

A typical accommodation schedule

Offices	*Toilets*	*Canteen and welfare*
Site manager	Canteen lady	Staff canteen
Surveyor	Staff	Labour canteen
Secretary	Clerk of Works	Drying room
General foremen	Site labour	First Aid room
Site engineer		Welfare equipment
Trades foremen		
Bonus surveyor		
Timekeeper		
Conference room		
Clerk of Works		
Sub-contractor's offices		

Preparation of Programmes

Introduction

There are a number of different types of programmes and also a number of different techniques which can be employed in their preparation and display. Different sizes and complexities of building projects warrant different kinds of programmes. It is therefore necessary to think carefully about the type of programme required, and who will be reading it. The programme is a pictorial expression of the planning process. It shows the inter-relationship between different operations and different tasks, when resources will be required on site and when they can be removed in the case of men, plant and equipment.

When the contract is underway, the progress made can be assessed against the plan (programme), and the effects of changed circumstances can therefore be measured. If the programme target times are realistic, they serve as a goal to site management to motivate them into action.

It is essential to number and date all programmes so that in the event of one being revised, everyone concerned can be made aware of the number of the current programme. Whilst this may sound like an admission of failure, it should be remembered that there are many variable factors completely outside the control of the contractor which may make adjustments necessary to the original programme. When changes do occur, without a programme to assess what the change does to the original plan, it can only lead to chaos. One justification for the programme then is that it provides a springboard for measuring the effects of proposed or actual changes.

Some of the variable factors.

1. Abnormally bad weather
2. Sub contractor failure to perform to standard
3. Trade dispute in the locality of the site
4. Trade dispute in another industry affecting the availability of material supplies or an essential service
5. Unpredictable labour shortage
6. Internationally, change in the availability of a material
7. Excessive variations to the contract
8. Overheating of the construction and materials supply industry due to action by H M Government
9. Inability of the consultants to supply information at the required time
10. Unforeseeable site conditions.

Some of these points might be foreseeable in part whilst others could be completely unpredictable.

Contractor performance

There are other reasons why a contract might fall behind programme, some of which could be related to poor performance by the contractor.

Amongst these can be identified:

1. Inefficient site management.
2. Bad planning and programming.
3. Lack of support at site management level of centrally conducted programming and control.
4. Misinterpretation of information provided in the documents, particularly with regard to quality standards.
5. Selection of wrong method or resources.

Types of programmes

There are five main types of programme, namely:

1. Pre-tender programmes.
2. Pre-construction programmes.
3. Overall programme.
4. Stage programmes.
5. Weekly programmes.

Pre-tender programme

Pre-tender programmes have already been considered in some detail. However, it is worth mentioning that one of the main purposes of this is to focus the minds of all concerned on a realistic programme by which the work can be carried out. If the architect accepts the programme, then he accepts the need to provide all the necessary information to match the programme; the same applies to his team of consultants.

To the contractor it provides a realistic basis to determine the cost and value of the work, and should his tender be successful, it provides the seed from which the pre-construction and overall programmes can germinate.

Pre-construction programme

When the contract has been won, further information will be available from the architect and his consultants. They will have continued working on the project during the pre-tender period as they were in no way dependent on any individual contractor's tender. Also, within the contractor's organisation resource availability may have changed from the forecasts which were made when the tender was being prepared.

Site manager involvement

It is vital that the site manager is involved in the development of the pre-construction programme. If there is little or no involvement then there will be little or no sense of commitment should construction fall behind programme. Any good site manager will make a serious attempt to retrieve lost ground if he has been responsible for the programme in the first instance. The programme chart should have provision for marking up progress as the work proceeds, see Figures 5.1 and 5.2.

Figure 5.1   Marking up progress

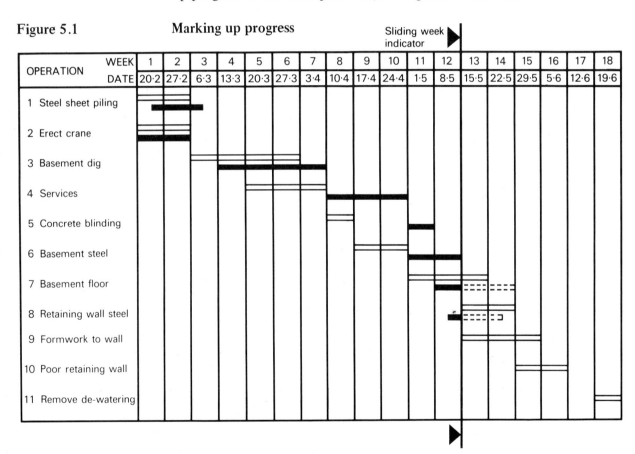

When work is not proceeding to programme and it is clear to the contractor that this is not just a small reverse trend which can be arrested and the contract quickly brought back to programme, the architect should be informed. If it becomes apparent after discussion with the architect that the situation cannot be retrieved, then material supplies, labour, sub-contractors, may require re-scheduling. In the case of mat-

Figure 5.2   An alternative method of marking up progress

Contract No.17 E.Lancs	Overall programme No2	prepared sms 7/5
		checked by R.H. 9/5

Month		June					July					August				Sept		
Wk/comm.		2	9	16	23	30	7	14	21	28	4	11	18	25	1	8	15	
Wk no.		25	26	27	28	29	30	31	32	33	34	35	36	37	38	39	40	

Location	Item No.	Descript.
Start area 7 Section 1 move westwards	1	Site scrape
Follow routing shown on site plan	2	Excavate foundations
Omit block 7 (Need to expose L.A.Sewer)	3	Concrete foundations
Block 8 & 9 first security plan for site entrance	4	Brickwork to d.p.c.
Follow routing on site plan	5	Services within wall lines
	6	Hardcore fill
	7	Concrete over site

Time Now →

Numeric progress figures marked on the bar chart (per week):

- Item 1 Site scrape: 45 | 45 50 | 95 50 145 50 195
- Item 2 Excavate foundations: 30 | 30 50 | 80 60 140 55 195 | 5 5 10 15 10 25 20 45 20 65 20 85 20 105 20 135 20 155 20 175 20 195
- Item 3 Concrete foundations: 2 2 14 16 20 36 25 61 30 91 34 125 31 156 30 186 | 5 5 10 15 10 25 20 45 20 65 20 85 20 105 20 135 20 155 20 175 20 195
- Item 4 Brickwork to d.p.c.: 4 4 10 14 20 34 20 54 20 74 29 94 20 114 | 4 4 10 14 20 34 20 54 20 74 20 94 20 114 20 134 20 154 21 175 20 195
- Item 5 Services within wall lines: 2 2 10 12 23 35 25 60 15 75 | 5 5 10 15 10 25 20 45 20 65 | 5 5 15 20 | 8 8 10 18 12 30 12 42 12 42 20 62 25 87 25 112 28 140 28 168 27 195
- Item 6 Hardcore fill: 8 8 10 18 12 30 | 8 8 10 18 12 30 12 42 20 62 25 87 25 112 28 140 28 168 27 195
- Item 7 Concrete over site: 8 8 10 18 | 8 8 10 18 12 30 12 42 20 62 25 87 25 112 28 140 28 168 27 195

erials, consideration must be given as to whether they can be received and stored rather than breaking faith with the suppliers. This will be measured against new problems of congestion, double handling, incr- eased wastage, security, vandalism, lock-up of capital, etc. Naturally, the longer the period of time given to the other parties concerned of an impending change of plan, the less disruptive it is.

Other kinds of programme.

In this chapter emphasis has been on production based programmes. However, it must be noted that there are other bases for programmes, particularly related to finance, ie cost, value, waste control. These are all dealt with elsewhere in the book.

The master or overall programme

Little need be said of the 'master', or as it is sometimes called, the 'overall programme'. The pre-construction programme is carried forward into the construction period as the master programme; it is adjusted and amended in the light of changing needs. It is most advisable that programmes are numbered consecutively as they are issued and as with superseded drawings, they should be carefully saved so that disagreements about payments, responsibility for delays, etc. can more easily be settled later. During the life of a contract, planning is an on-going management activity. All contracts managers hope that the master programme will not require amending, but for many it is a dream, not a reality. Ob- viously, short delays would not justify altering the overall programme.

Stage programme

As the name suggests, a programme for a stage or part of the total works. The duration of such a programme will be varied to suit the particular needs of the job and sometimes advisably varied in duration on the one job. For example, on a concrete framed building with cladding panels, it may be found that one storey of frame can be erected in five weeks. After the erection of the frame, no single operation dominates the production. In these circumstances, a five week stage programme might be appropriate for the frame erection and the remainder of the works carried out with programme duration coinciding with interim certificate periods.

On a large speculative housing development, it may be advisable to have detailed planning (stage programme) ahead for four weeks; after the elapse of one week, a new week is 'ghosted' on to the end of the programme, thus renewing the four weeks advance planning.

Probably the most common arrangement is to have programme dur- ations coinciding with interim certificates. It is obvious that as soon as one programme is embarked upon, planning and scheduling of labour, sub contractor, materials, and plant must start for the next programme.

Preparation of the overall programme

It is usually a joint responsibility of the contracts manager and site manager to be looking ahead to the needs of the contract. In general, the contracts manager sees this job as part of the total work in progress of the firm, attempting to utilise the firm's resources to the fullest extent, satisfying the client, and in so doing enhancing the firm's reputation. The site manager views a narrower, more detailed horizon, one in which he identifies constructional or labour problems and looks for their solu- tion. They will discuss the strategy and tactics of the situation, but the contracts manager, carrying the greater responsibility, will have the final say in decision making.

Once they have forecast the production to be carried out in the period of four to eight weeks ahead, the site manager can instruct a planner on the preparation of the next stage programme. This may involve a planner visiting site for two or three days per month. A large site may be able to fully use the services of a resident planner.

Basic framework of a stage programme.

The time scale should be in half day units and the work description should be individual trade activities. The main aims of the programme are to provide general foremen, trades foremen (including here sub-contractor's foremen) with an immediate reference point which will show them where they should be working, when they should finish, which other trades are dependent on them finishing on time, and, what work they will proceed to when the current tasks are completed. Apart from this, the programme is a valuable aid to the site supervisors in planning the use of space, material stocking, and the use of labour. Finally, it helps in measuring progress of the work.

Weekly programmes

Once a week the site manager or general foreman will hold a meeting with craft and sub-contractor foremen in attendance to plan the work for the coming week. This meeting will magnify the detail of the stage programme to an hourly basis.

The meeting will normally be held on Thursday afternoon or Friday morning when the production for the week in progress can be accurately assessed. The meeting involves each foreman in the planning process and as such develops motivation and a sense of commitment.

The work planned can be presented in Gantt Chart form with individual trade work schedules derived from it. If ready-mixed concrete is being used, then daily schedules for concrete can easily be extracted. Where resources such as a fork lift truck or tower crane are being shared by the gangs, daily plant programmes can be developed from the weekly plan.

Drafting and adjusting

It is unlikely that the first attempt at drawing up the programme will prove satisfactory. Typical needs for re-adjustment are as follows:

1. The programme may 'over run' the contract completion date.

2. Key resources such as a tower crane or groups of workmen may be shown in use before they will be available to the site.

3. Operations may be shown in progress at times of the year which are clearly disadvantageous, eg external painting in January.

Schedules which can be derived from the programme

1. Information required schedule

2. Plant schedule

3. Labour schedule (directly employed)

4. Sub-contractor schedule

5. Material schedule

6. Equipment schedule

7. Scaffolding schedule

8. Accommodation schedule

Features of a good overall programme

1. The programme should be financially sound. The objective should be to reach a quick build up of activity and to maintain this level

Figure 5.3

Phased completion

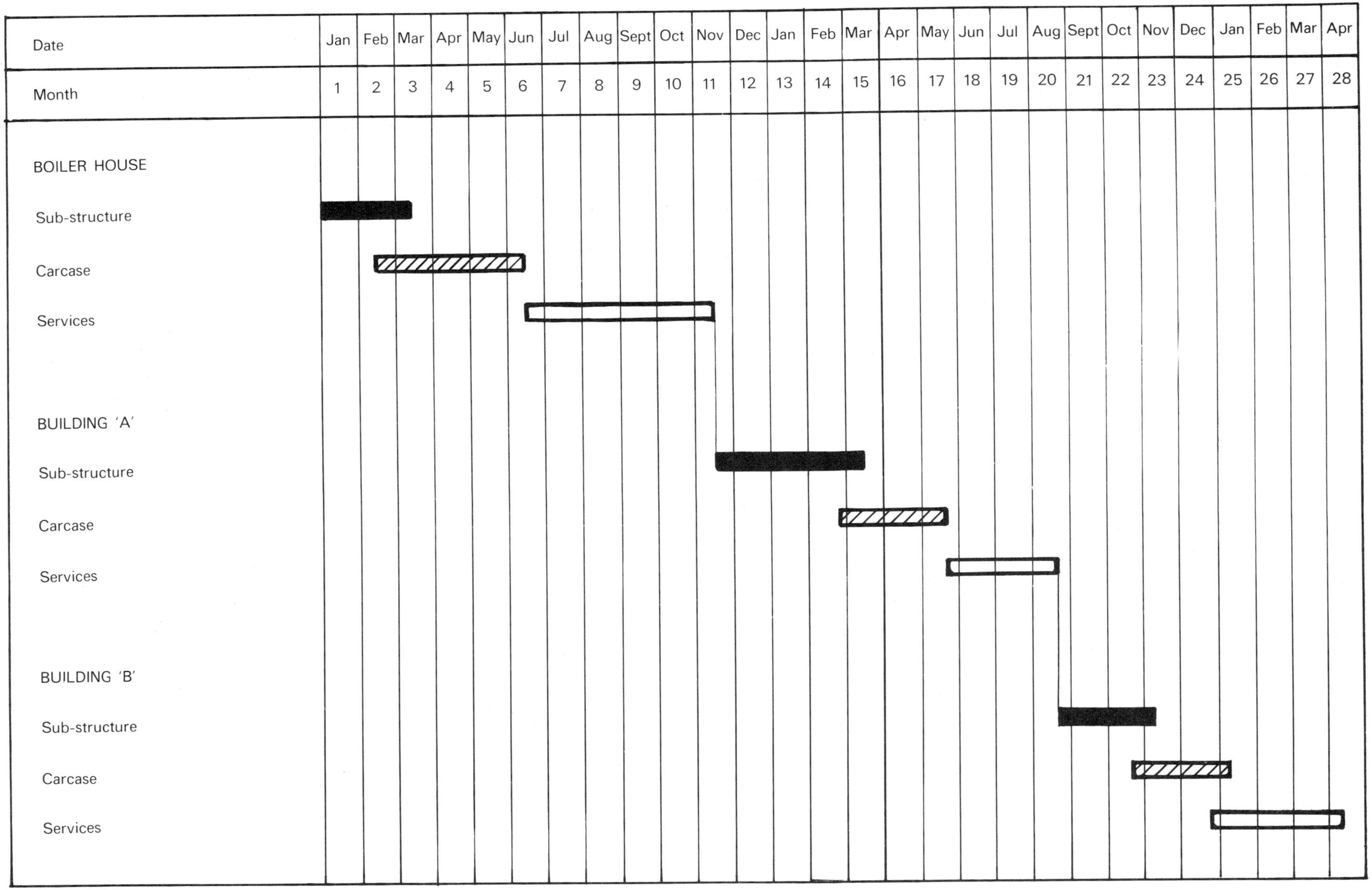

Date	Jan	Feb	Mar	Apr	May	Jun	Jul	Aug	Sept	Oct	Nov	Dec	Jan	Feb	Mar	Apr	May	Jun	Jul	Aug	Sept	Oct	Nov	Dec	Jan	Feb	Mar	Apr
Month	1	2	3	4	5	6	7	8	9	10	11	12	13	14	15	16	17	18	19	20	21	22	23	24	25	26	27	28

BOILER HOUSE
Sub-structure
Carcase
Services

BUILDING 'A'
Sub-structure
Carcase
Services

BUILDING 'B'
Sub-structure
Carcase
Services

of activity with a steep fall off at the end of the job. This ensures the setting and use of realistic levels of site supervision and an efficient use of overheads.

2. The construction programme should, wherever possible, give continuity of employment for a trade group once they arrive on site. Intermittent operations create problems of what to do with the labour when that item of work is stopped.

3. If the site is to be closed for holidays, these times should be indicated. Also, significant milestones should have attention drawn to them by means of the symbol ∇

   Examples of these are:

   (a) All work up to dpc completed.

   (b) Boiler house operational.

   (c) Building 'x' watertight.

4. The operations are numbered consecutively and listed on the left hand side of the special programme form. It is usual to write them in earliest commencing date order. If a programme contains more than twenty or so operations, each operation bar should be numbered on the right to minimise the danger of misreading.

5. An overall programme will show major operations and not minor activities.

6. The key operations or cycles should be readily identifiable. The 'leads' and 'lags' between preceding and succeeding operations should be assessed.

7. Calendar dates and contractual weeks or months should be shown. If the duration of the contract is less than eighteen months, a week is the most usual time unit. However, if the contract runs for more than a year and a half a month is the most convenient time unit.

8. If it is feasible to complete the contract in a shorter period than that allowed in the Appendix to the Contract, the contract's manager will seek approval from the client via the architect to plan to this end. The saving in site overheads represents a second line of profit.

When programming the erection of three buildings of a similar type of construction on the same site, is it better to erect each building separately, taking building number one from foundation to roof before starting building two, or to optimise on utilisation of labour? Figure 5.3 shows a concentration on one building at a time whilst Figure 5.4 demonstrates the continuous use of the labour once it is on site.

Features of the two programmes

	Figure 5.3	*Figure 5.4*
Hand over date for boiler house	Month 11	Month 11
Hand over date for building A	Month 20	Month 14
Hand over date for building B	Month 28	Month 17
Contract duration	28 months	17 months

From the client's viewpoint, it may be preferred to have the phased hand over illustrated in Figure 5.3. From the contractor's viewpoint, Figure 5.4 gives continuity of work for the labour force, a lower cost in site preliminaries and a quicker return on the capital employed.

Figure 5.4

Continuous use of labour

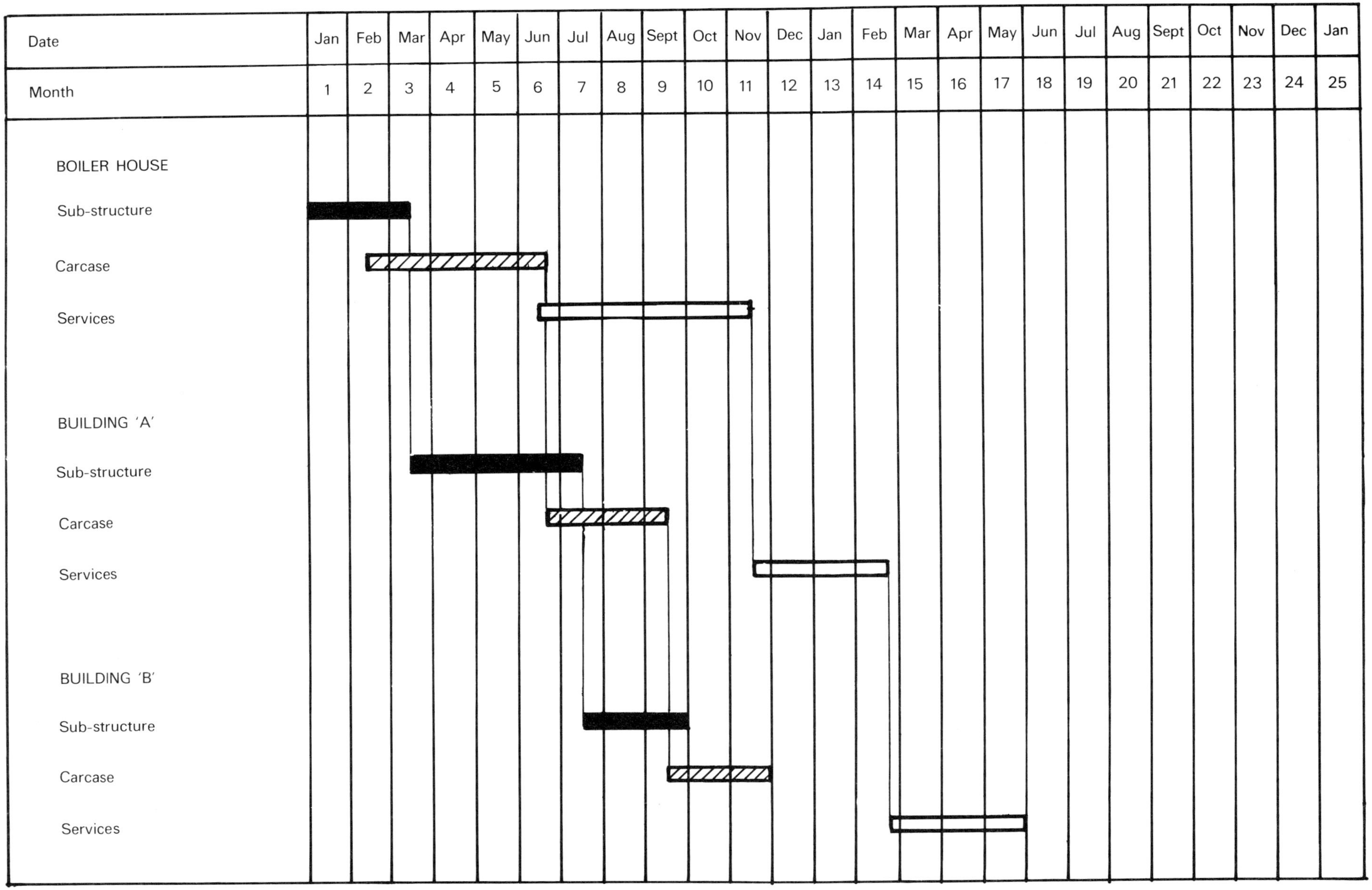

Date	Jan	Feb	Mar	Apr	May	Jun	Jul	Aug	Sept	Oct	Nov	Dec	Jan	Feb	Mar	Apr	May	Jun	Jul	Aug	Sept	Oct	Nov	Dec	Jan
Month	1	2	3	4	5	6	7	8	9	10	11	12	13	14	15	16	17	18	19	20	21	22	23	24	25

BOILER HOUSE
Sub-structure
Carcase
Services

BUILDING 'A'
Sub-structure
Carcase
Services

BUILDING 'B'
Sub-structure
Carcase
Services

HOW TO PREPARE A GANTT CHART

There are a number of planning techniques in use but the Gantt chart method of display is still the most popular in the construction industry. Gantt charts can be used for overall programmes or weekly programmes. The following information is required before commencing:

1. The relevant construction drawings.

2. A method statement for major and intricate operations.

3. A schedule of bulk quantities in operation form.

4. Anticipated performance rates - these would be by estimate or from historical records.

5. Notification of any time restrictions such as when key labour, resources, or materials, become available.

It is also necessary to have a suitable programme form and some programme calculation sheets. It is usually more convenient to draft the programme in pencil before finally inking in.

Drainage example

Figure 5.5 shows the site plan for a proposed sewer for a local authority. Assume that the planner has been asked to prepare a pre-tender programme which will be submitted with the tender to the local authority.

Information available

The site will be available to the contractor on Monday, May 2nd and the whole of the work must be completed by Friday, August 19th, a period of 16 weeks.

The Estimator's Site Investigation Report

Site description

The site is very flat with a slow rise from the Disposal Works to Manor Close. The area is well drained with agricultural ditches. The top soil is a loam and overlies a medium hard chalk. The depth of the top soil varies between 300 mm and 1000 mm. The site is not liable to flooding. There are some signs of water-logged ground, commencing immediately south of the railway embankment and continuing for about 130 metres southwards.

Roads

Chatterton Lane has a good carriageway and should present no problems. Dayspring Lane is also metalled but is generally in much inferior condition compared to Chatterton Lane. The verges of Dayspring Lane are very weak and wet, and the carriageway is only four metres wide. Vale Road and Tanners Road are both in good condition.

Services

There is a metered supply of water available at Manor Nursery. The Contract Manager confirms that an electricity supply is not required. There is a public telephone kiosk in Manor Close. (The reader should note that on a project of this type it would not be worthwhile installing a telephone, since there would be no need to man a site office and a yard bell would not be audible for more than 80 metres or so.)

Disposal of spoil

Arrangements can be made for free disposal on Messrs. Fosters Brewery site which is located one mile from the Disposal Works in the direction of Chatterton.

Figure 5.5   **Site plan Ewood drainage**

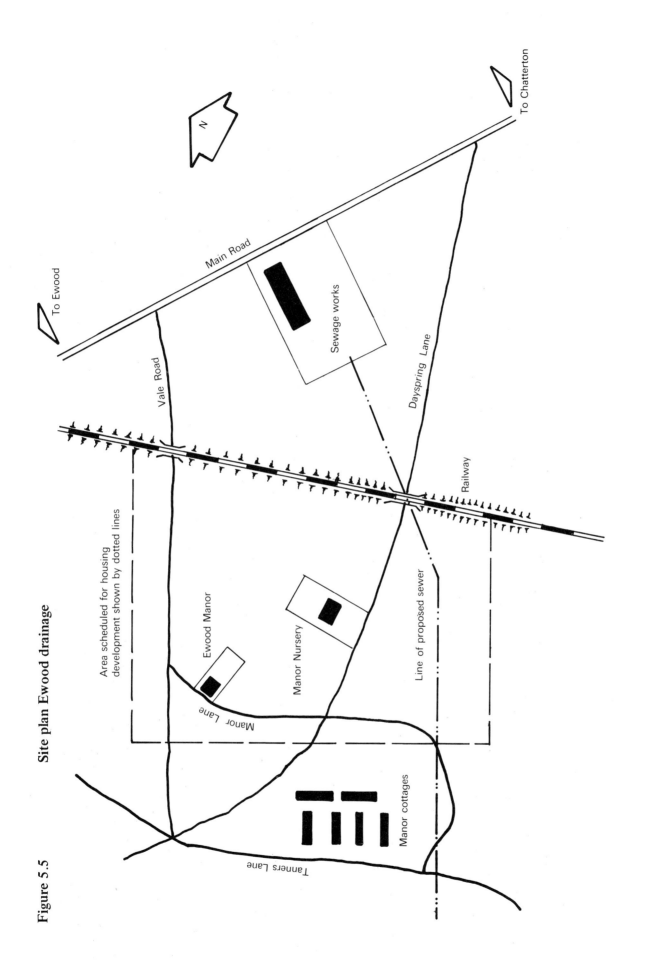

Storage

Manor Nursery is not cultivated because the owner has retired. Valuable storage buildings can be rented for a total sum of £100 for the contract period. Owner E. Smithers, 4 Dayspring Lane (no telephone).

Fences, etc.

Eleven hedges cross the line of the sewer. Cattle will be using the field containing inspection chambers Nos 4 to 13 inclusive, hence temporary fencing will be needed to contain them.

Temporary closure of Dayspring Lane

Discussions have already taken place between the Borough Surveyor and the police concerning the closure of Dayspring Lane. Agreement has been reached, in principle, on a seven day closure, subject to the contractor providing and fixing adequate signs for the proper re-routing of traffic. (Signs to be approved by the police). Seven days notice of closure to be given. New temporary route to be effective from 05.00 hours on a Monday morning. The change-back to the normal route to be effected at the same hour one week later.

End of Report   Signed: S.M.Smith
                  Date: 13th March

Preliminary notes to the Method Statement

1. A long run of trenching has to be dug. The work content is almost exclusively repetitive.

2. Drainage pipes have to be carried over rough terrain to the trench. The pipes are bulky and require care in carriage to avoid damage.

3. A large quantity of gravel must be transported to the trench side and later trickled into the trench, to form first the bed and later the surround.

4. The pipes have to be lowered into the trench and laid.

5. Pre-cast concrete manhole sections and small quantities of mixed concrete must be transported.

6. In essence, this contract contains six major activities, each having a fairly considerable work content.

   (a) The trench dig

   (b) Laying the loose fill

   (c) Laying the pipes

   (d) Building the inspection chambers

   (e) Completing the loose fill surround

   (f) Back filling the trench

7. If these six activities can be set up with an equal work rate, ie, balanced gangs, there is a good chance of attaining high productivity and a successful financial conclusion. The relatively minor tasks should be made to fit in with the six major ones.

   Assuming that the objective is to complete the job to the agreed specification at the least cost, each of the major activities will be considered in turn to find the cheapest means of execution.

8. Various types of digging machines should be compared. These could include a bucket chain trencher and a conventional hydraulic digger. The comparative studies should include all the ancillary costs such as transport of machine to and from the site, fuel, tracks, if any, etc. The terms of hire daily rate or weekly and the total real cost of hiring should be compared.

Method Statement

Digging

Main trench digging will be carried out with a bucket chain trenching machine.

The extra width for manholes and digging out in the Dayspring Road section will be by hydraulic digger. Any rough spots in the main trench dig will also be done by the hydraulic digger.

Loose fill

This will be delivered to the sides of the trench in company vehicles and trickled into the trench by a Drott shovel.

Space

One side of the trench will be reserved for spoil storage and getting spoil away. The other side for the receipt, storage and handling of materials and carrying out drainage operations.

Concrete

Concrete for inspection chambers will be site mixed and transported by site dumper.

Handling drain pipes

These will be handled across the site by fork-lift truck. The sub-contract drainlayers will manhandle them into the trench using rope slings (weight of each pipe 40 kilos).

Handling inspection chamber sections.

These will be carried across the site by fork-lift truck and lowered into position by means of a small tractor-mounted crane.

Backfilling

This will be carried out with the Drott shovel and hydraulic digger.

The operation number is the sequence by which jobs are commenced and the order in which they will appear on the Gantt chart (Table 5.1). The quantities shown in column 3 are expressed in operation form, thus in operation number 4 pipes are enumerated. This can quickly be converted into lorry loads. The estimated performance rate is given in column 6. This would be derived from historical records or by analytical estimating. By dividing the quantity in column 3 by the estimated performance rate the total hours shown in column 7 can be determined.

TLO in column 8 stands for tradesmen, labourers, and others. Operators, banksmen, chainmen would make up the main bulk of the others. By dividing the total hours in column 7 by the number of producers the programme weeks can be found, eg, operation 6 contains 720 hours work for four drainlayers. That is, $720 \div 4 = 180$ hours work each or 4½ programme weeks.

Programme

The programme (Figure 5.6) can now be drawn up from Table 5.1. It is also possible to extract from Figure 5.6 a labour summary Table 5.2

Table 5.1   Programme calculation sheet

Operation Number	Operation	Quantity	Unit	Trade	Estimated performance rate	Total hours	Gang size TLO	Programme weeks	Remarks
1	Erect compound	—	—	Carpenter	—	150	2 2	1	
2	Set out			Engineers		240	2 2	3	
3	Excavate trench	3,600	m	s/c Excavator		220	2 1	5	1 lorry for spoil to tip
4	Materials–pipes handling fill manhole sections	2,400 / 1,050 / 144	No / m³ / No	Excavator	Intermittent deliveries	360	2 2	9	Drott shovel and fork lift truck.
5	Loose fill bed	3,600	m	Excavator	8 per hour	450	3 1	4	Drott
6	Lay sewer	3,600	m	Drainlayer	4 per hour	900	4 4	6	Includes testing
7	Dayspring Lane	—	—	—	—	—	—	—	See detailed programme
8	Loose fill surround	3,600	m	Excavator	6 per hour	600	3 1	9	Intermittent
9	Inspection Chambers	36	No	Concrete	5 hours each for the team	180	3 1	6	Intermittent
10	Back fill	3,600	m	Excavator	12 per hour	320	4	9½	Drott Intermittent
11	Reinstatment of fences, etc.							3	
(1)	(2)	(3)	(4)	(5)	(6)	(7)	(8)	(9)	(10)

Figure 5.6   Programme Ewood drainage

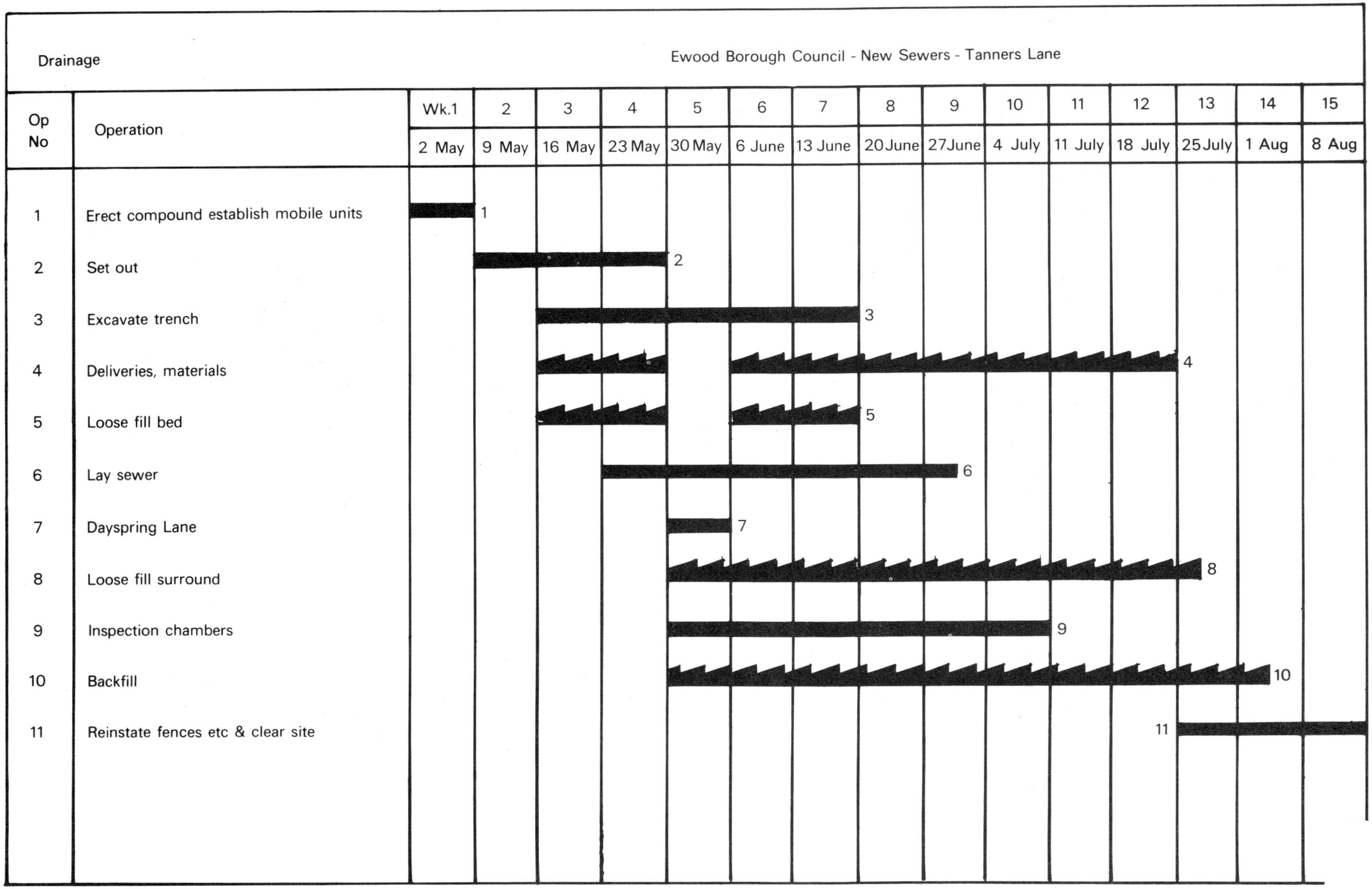

Ewood Borough Council - New Sewers - Tanners Lane

Drainage

Op No	Operation	Wk.1 2 May	2 9 May	3 16 May	4 23 May	5 30 May	6 6 June	7 13 June	8 20 June	9 27 June	10 4 July	11 11 July	12 18 July	13 25 July	14 1 Aug	15 8 Aug
1	Erect compound establish mobile units	1														
2	Set out		2													
3	Excavate trench							3								
4	Deliveries, materials												4			
5	Loose fill bed								5							
6	Lay sewer									6						
7	Dayspring Lane						7									
8	Loose fill surround													8		
9	Inspection chambers											9				
10	Backfill														10	
11	Reinstate fences etc & clear site												11			

81

and a plant summary Table 5.3. It should be noted that no details are given of operation 7 as this is a complex operation involving a temporary road closure and breaking into a metalled road, laying the sewer and the reinstatement of the road. This would be discussed in detail and a separate programme of one week's duration developed for the control of this work.

Digging

The engineer would try to maximise on the use of the Allen trencher and would use the Camberley digger in the main trench for any rough spots such as boulders or difficult trees to aboid delaying the trenches. The trencher would also 'pass straight through' the inspection chambers, the increased width for these would be achieved by the Camberley. The Camberley would also be responsible for the work in Dayspring Lane.

Spoil

If 1050 m³ of loose fill is to be imported and also 2400 pipes, it follows that a considerable amount of spoil must either be 'spread and levelled' or carted away. If the Bill calls for it to be carted away and assuming that a spoil truck can operate up to the trench area, then a spoil truck can be loaded directly by the trencher as it digs. To remove, say, 1500 m³ in an 8 m³ truck will take approximately 200 loads. Assuming a loading time of 20 minutes and a return trip to the tip of 40 minutes, the truck would take 320 m³ to the tip per week (40 hrs x 8 m³). The truck would be fully engaged on this work for 5 weeks.

Ground water

It would be advisable to have a pump in attendance so that in the event of water seeping into the trench, it can be dealt with quickly. Naturally one would hope that it would not be required.

Timbering to the trench

The drainlaying sub-contractor has contracted to provide trench side supports. His gang of 8 men will carry the timbering forward so that the trench is always properly supported in the sections in which they are working. At any one time they would have four sections of drain being laid.

Table 5.2 **Labour summary**

Category	No.	Person if known	Date on-site	off-site
Engineer-in-charge	1	R. Brown	2/5	25/7
Engineers	2	J. Gray, M. Williams	9/5	27/5
Chainmen	2	— —	9/5	27/5
Carpenters	2	— —	2/5	6/5
Labourers	6		2/5	12/8
Lorry driver	1		16/5	17/6
			1/8	12/8
Sub-contractors				
Allen Trencher plus operator	1	X Y Plant	16/5	17/6
Drainlayers	8	R. Nolan Ltd.	23/5	30/6
Camberley digger	1	X Y Plant	16/5	5/8
Fork lift truck	1	X Y Plant	30/5	24/6
Drott	1	Gold Plant	16/5	12/8

Figure 3.7   A typical stage programme

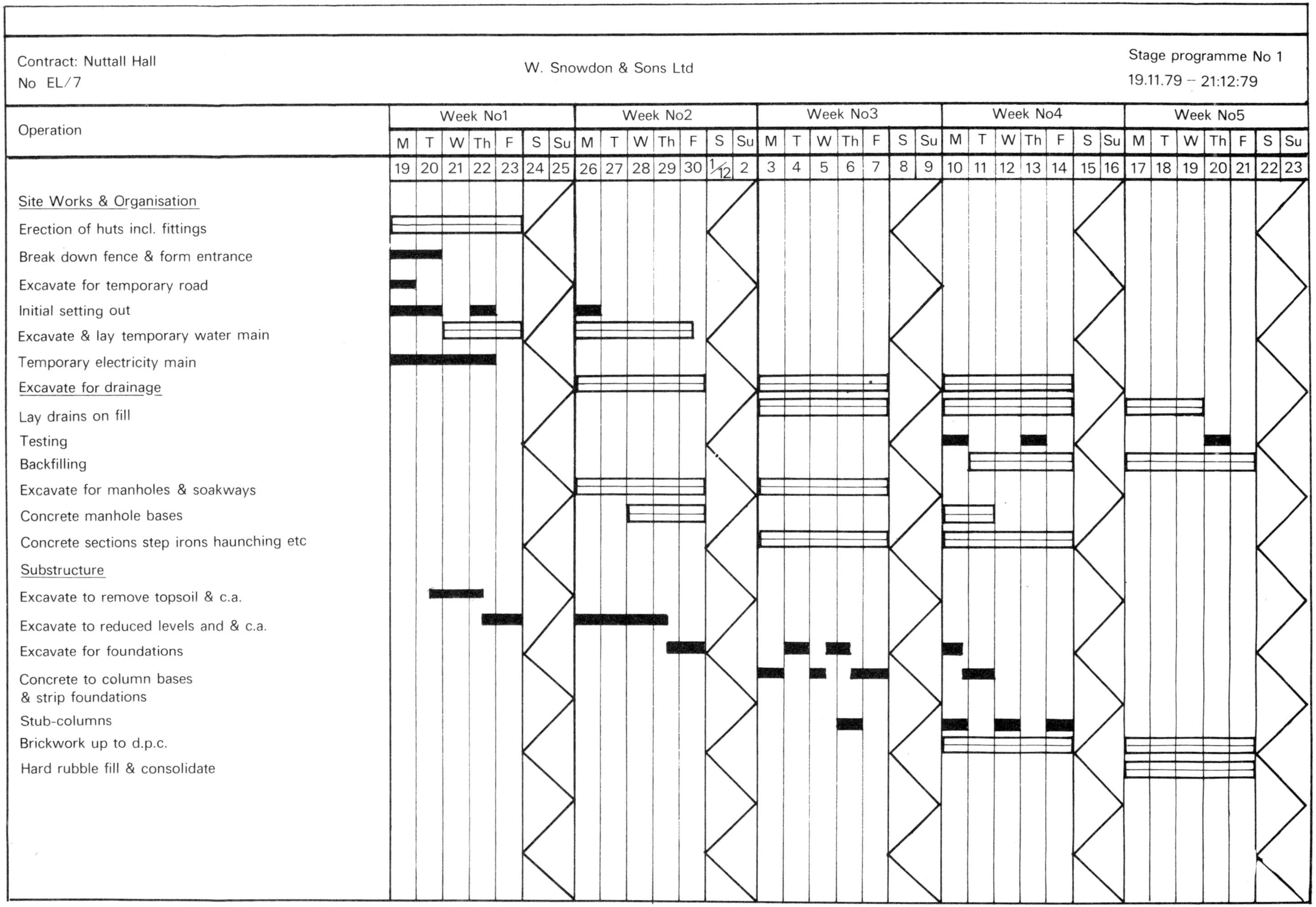

Contract: Nuttall Hall
No EL/7

W. Snowdon & Sons Ltd

Stage programme No 1
19.11.79 – 21.12.79

Operation	Week No1								Week No2								Week No3								Week No4								Week No5							
	M	T	W	Th	F	S	Su	M	T	W	Th	F	S	Su	M	T	W	Th	F	S	Su	M	T	W	Th	F	S	Su	M	T	W	Th	F	S	Su						
	19	20	21	22	23	24	25	26	27	28	29	30	1	2	3	4	5	6	7	8	9	10	11	12	13	14	15	16	17	18	19	20	21	22	23						

Site Works & Organisation

Erection of huts incl. fittings

Break down fence & form entrance

Excavate for temporary road

Initial setting out

Excavate & lay temporary water main

Temporary electricity main

Excavate for drainage

Lay drains on fill

Testing

Backfilling

Excavate for manholes & soakways

Concrete manhole bases

Concrete sections step irons haunching etc

Substructure

Excavate to remove topsoil & c.a.

Excavate to reduced levels and & c.a.

Excavate for foundations

Concrete to column bases
& strip foundations

Stub-columns

Brickwork up to d.p.c.

Hard rubble fill & consolidate

83

Table 5.3 **Plant summary**

Operation number	Machine	Size	Period (weeks)	Date From	To	Hire rate	Total Planned Cost £	Remarks
3	Allen trencher	16/60	5	16/5	17/6	xxx	xx	Machine to be off loaded Four sets of matting to be off loaded — ······ Fuel ······
3,4,5 7,8,10	Camberley digger Hydraulic back hoe	0.45(18)	12 12	16/5	5/8	xx	xx	To be unloaded at ······ Ancillary equipment as follows ······
4,9	Fork Lift truck		4	30/5	24/6	xx	xx	Deliveries of pipes and manhole rings to be planned accordingly.
4,5,8,10	Drott		13	16/5	12/8	xx	xx	
3,10	Lorry	8m³	5 2	16/5 1/8	17/6 12/8	x	x	
1	Compressor		1	31/5	4/6	x	x	
1	Pump	100mm	12	23/5	12/8	x	x	
(1)	(2)	(3)	(4)	(5)		(6)	(7)	(8)

THE CRITICAL PATH METHOD

This is a system of planning which was developed in the United States of America in the 1950's. Critical Path is one of a number of network analysis techniques. When it was introduced into this country in the 1960's, many of the major construction companies were quick to use it however, at the moment there is perhaps some disenchantment with the system. The name critical path grows out of a unique advantage of the method. It shows graphically which jobs, and they are seldom more than 10% of the total, are critical, that is, will delay the whole project if they are delayed. These are the jobs which must be carefully controlled and speeded up if it is necessary to shorten the operation time.

Any project can be divided into a number of separate jobs. In Critical Path Method these separate jobs are called activities.

Activities

An activity is represented by a single arrow, the length of which is of no significance, activities flow from left to right. In the examples which follow, letters are used to represent jobs of work.

Example 5.1

Activity C must follow A

Solution

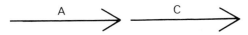

Example 5.2

Activity Y must follow X and Z

Solution

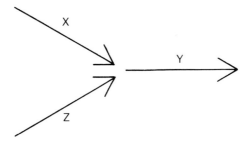

Example 5.3

Activities C and D must follow L and P

Solution

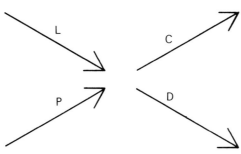

Example 5.4

Activity A follows after activities X and Y, and D and E both follow after A.

Solution

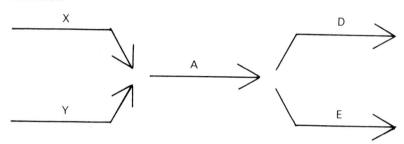

Example 5.5

Activity P follows after R and C, R and E both follow after L

Solution

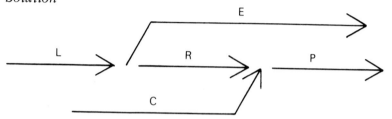

Example 5.6

Activity D cannot start until activities A and B are completed. Activity C cannot start until A is completed, but note that C is independent of both B and D. *First attempt*

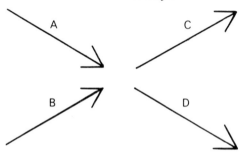

This is incorrect because C is shown to be dependent on B. Sometimes it is difficult to express the true logic. In these cases 'dummy' activities can be used. A dummy activity is represented by a dotted line.

Correct solution

C is shown independent of both B and D.

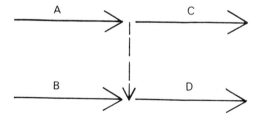

Example 5.7

A planner's first draft of a project may be as follows:

86

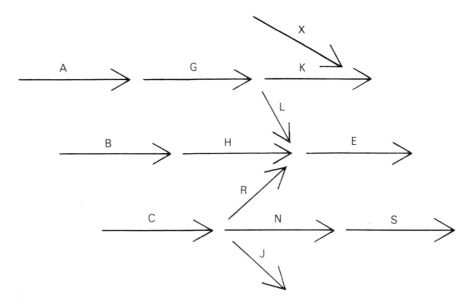

The diagram should now be tidied up. There should be only one start point and one end point. Where an activity has no logical prior activity, it should be linked back to the start. Where an activity has no logical succeeding activity, it must be linked forward to the end.

Example 5.7 Tidied up

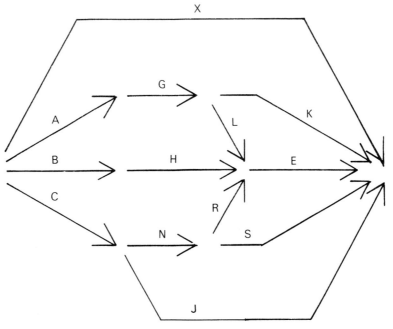

Activity Numbering

So far the examples have been concerned with expressing the logic of situations. The next stage is to consider activity numbering. This is necessary so that advantage may be taken of the computer. The numbers permit each activity to be recognised by a unique pair of numbers.

Consider Example 5.6 again.

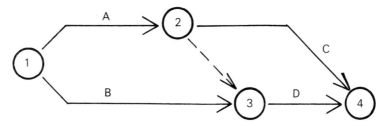

87

Activity	Recognised by
A	1.2
B	1.3
Dummy	2.3
C	2.4
D	3.4

The dummy was used in Example 5.6 to help to express the true logic of the situation. Dummies are sometimes necessary to preserve the rule whereby each activity is recognised by a unique pair of numbers.

Example 5.8

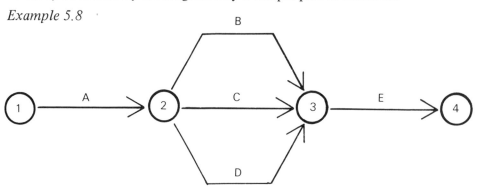

This logic shows that when activity A has been completed, activities B, C, and D can start. When these three have been completed, then E can start. If numbers are inserted at the ends of the activities, then B, C, D will be seen to have the same numbers. To rectify this, the diagram is redrawn as follows:

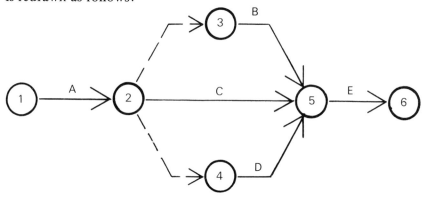

Each activity now possesses a unique pair of numbers.

The number at the head of an activity should always be larger than the number at the tail. This is a general rule which sometimes has to be broken, for example, when drawing sub-networks.

Time-based dummies

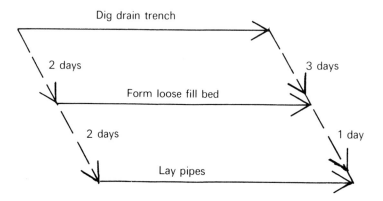

In laying a long drain, it would not be unusual to start preparing the loose fill bed for the pipes after the digging operation has a good 'lead'. Similarly, once the loose fill is well under way the laying of drain-pipes could start. To express the extent of the delay between the start time of each of these three operations, time-based dummies are used. The logic expressed is 'Dig drain trench' starts two days before 'form loose fill bed'. 'Form loose fill bed' starts two days before 'lay pipes'. 'Dig drain trench' must finish three days before 'lay loose fill bed'. 'Loose fill bed' finishes one day before 'lay pipes".

Scheduling – Events

An event may be defined as a stage in a project which marks the completion of one or more activities or the start of one or more activities. The numbered circles on the diagram indicate the event number. When the anticipated duration of an activity has been agreed, it is entered on the diagram against the activity. There are two 'times' associated with an event. One is the earliest event time (EET), the other the latest event time (LET). The earliest event time is the longest timewise chain leading into an event.

To calculate EET travel from left to right on the diagram and put the figures in the left hand box.

Example 5.9

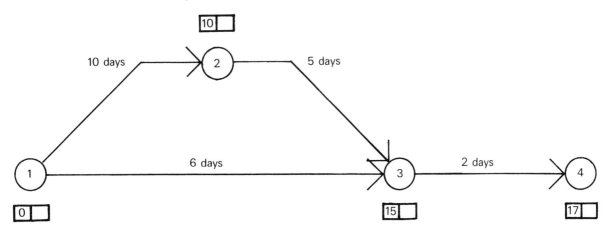

There are two chains leading into event 3, namely:

1. 10 + 5 days = 15 days

2.   6 days

Event 3 will not be reached until 15 days

All durations must be in the same time units, ie, days or hours, but not a mixture of the two.

The latest event time (LET) is the latest time by which an event must be reached, that is, all the activities leading up to that point must be completed in order that the whole project can be finished by the earliest project completion day. LET's are calculated from the end of the project back to the beginning.

To complete Example 5.9

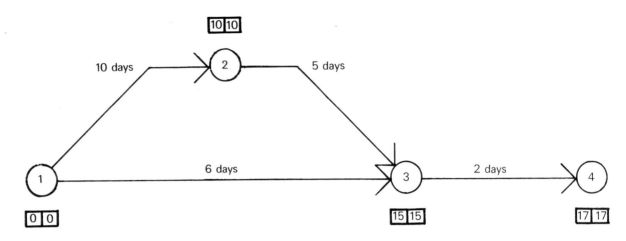

The LET at Event 4 is 17
                at Event 3: 17 - 2 = 15
                at Event 2: 15 - 5 = 10
                at Event 1 two chains must be considered:
                        10 - 10 = 0
                        15 - 6 = 9.
                The smallest number is selected.

Table 5.4   **Schedule of activities for Example 5.10**

Activities	Duration in days	Start Earliest	Start Latest	Finish Earliest	Finish Latest	Total Float
1 — 2	5	0			5	
1 — 3	3	0			17	
2 — 3	12	5			17	
2 — 4	6	5	23	11	29	18
3 — 4	7	17			29	
3 — 5	15	17			32	
4 — 5	3	24			32	
4 — 6	5	24			42	
5 — 6	10	32			42	
(1)	(2)	(3)	(4)	(5)	(6)	(7)

In column one activities are listed in least 'tail' number order:

                Tail ⟶ Head.

The Critical Path

Where there is coincidence between the EET and LET at an event point it indicates that this is a critical event, that is, the critical path passes through this point. In the case of Example 5.8 all events are seen to be critical, but not all activities are critical. In this example the following activities make up the critical path.

Activities 1 - 2 + 2 - 3 + 3 - 4

Durations in days 10 + 5 + 2 = 17 days

If any one of these activities is delayed, the completion date for the project will be delayed. Activity 1 - 3 is not critical. There is a period of 15 days when this 6 day activity can be accomplished, thus there are 9 days of float (sometimes called slack) associated with this activity.

Preparing the schedule of activities.

So far, calculations have been concerned with earliest and latest start times for events. The next step is to calculate the earliest and latest start for each activity. When this is known, the critical activities can be identified.

Example 5.10

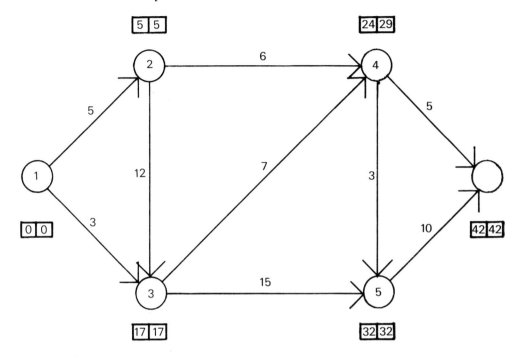

The numbers in column 3 are derived from the EET at the tail event. The numbers in column 6 are derived from the LET at the head of the activity. Now consider activity 2 - 4. This 6 day activity can commence as early as day 5, its earliest finish time is therefore:

day 5 + duration of activity
= 5 + 6 = 11 (see column 5).

Its latest start time is found by subtracting the activity duration from its latest finish time, ie:

29 - 6 = day 23 (see column 4).

There is a period of 24 days (between day 5 and day 29) when this 6 day activity can be carried out; therefore this activity has 18 days of float. Once the true meaning of float is understood the easiest method of calculating it is either by:

(a) subtracting the earliest start from the latest start, or

(b) subtracting the earliest finish from the latest finish, ie, activity 4 - 6:
34 - 24 = 13 days float
42 - 29 = 13 days float.

It should be noted that it is usual to show numbers at each event point as follows. The method used earliest of placing numbers in boxes was done solely for reason of clarity.

Figure 5.8

Event time

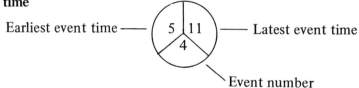

Earliest event time ——— (5 | 11) ——— Latest event time

4 — Event number

Time available for activity - activity duration = Total float.

24 - 6 = 18 days (see column 7)

The reader is advised to make a copy of Table 5.4 and attempt to fill in the missing numbers which can be checked with Table 5.5

Table 5.5

Example 5.10 completed

Activities	Duration in days	Start Earliest	Start Latest	Finish Earliest	Finish Latest	Total Float
1 — 2	5	0	0	5	5	0
1 — 3	3	0	14	3	17	14
2 — 3	12	5	5	17	17	0
2 — 4	6	5	23	11	29	18
3 — 4	7	17	22	24	29	5
3 — 5	15	17	17	32	32	0
4 — 5	5	24	37	29	42	13
5 — 6	10	32	32	42	42	0

It should be noted that four activities have no float. These are the critical activities and together form the critical path. The critical path may be shown in red on the network diagram or two lines across the tail of each critical activity.

Updating the programme

As work is being carried out, reference will be made to the network to measure and record actual progress against planned progress. The frequency with which this is done depends very much on circumstances. Where work is falling behind programme, a manager has to decide what corrective action to take in drastic situations. It may mean revising the logic, adding new activities, or combining old ones.

Three main aspects

The three main aspects of updating are:

1. To 'line thro' those activities which have been completed.

2. Estimate the time for completion of those activities in progress.

3. In the light of experience revise any duration estimates for future activities.

When the network has been modified in accordance with these points, the last act is to re-calculate the event times either to the end of the contract, or to such a point in the contract, when it is expected to resume the original time estimates.

Two programmes
It is advisable to have two copies of the programme displayed, one which remains unaltered throughout the progress of the work. This is retained for reference purposes. On the copy which is to be amended, all completed activities should be indicated by 'blocking in' or hatching through the arrows and setting the activity durations to zero. If an event has been achieved, the event should be hatched through and the contract day/week on which the event was achieved indicated.

Figure 5.9

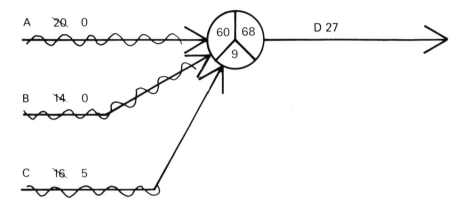

Figure 5.10

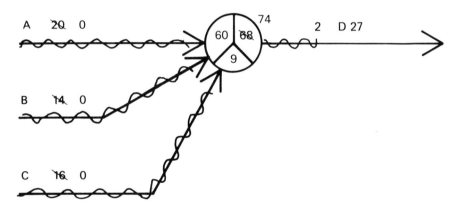

Figure 5.9 shows that both activities A and B have been completed. Activity C is not complete. It is estimated that it will take five days to complete.

Figure 5.10 shows that activities A,B and C have all been completed. Event number nine was arrived at on day seventy-four; two days' work has been accomplished on activity D. Updating should be done in a different colour each time and a reference made to this somewhere on the diagram, viz:

Green   16 April
Red   14 May
Blue   11 June
Green   9 July, etc.

The classification of float

There are different classes of float and whether it is very significant is difficult to conclude. For a lot of work it is sufficient to know that there is total float and to level resources within the confines of total float.

How to calculate float

Total float

Latest date of succeeding event minus earliest date of preceding event minus the activity duration.

Free float - early

Earliest date of the succeeding event minus earlier date of the preceding event minus the activity duration.

Free float - late

Latest date for succeeding event minus latest date for preceding event minus the activity duration

Example 5.11

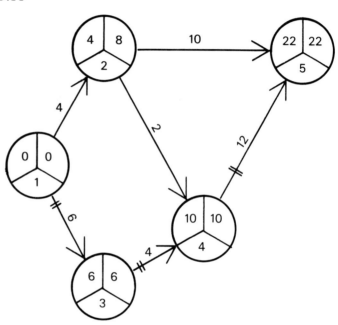

Table 5.6   **Activities schedule Example 5.11**

Activity	Duration	Start		Finish		Total Float	Free Float Early	Free Float Late
		Earliest	Latest	Earliest	Latest			
1 — 2	4	0	4	4	8	4	0	4
1 — 3	6	0	0	6	6	0	0	0
2 — 4	2	4	8	6	10	4	4	0
2 — 5	10	4	12	14	22	8	8	4
3 — 4	4	6	6	10	10	0	0	0
4 — 5	12	10	10	22	22	0	0	0

Consider Activity 2 - 5

Total float = latest date of succeeding event − earliest date of preceding event + duration

= 22 - (4 + 10)

= 8.

Free float early = earliest date of succeeding event − earliest date of preceding event + duration

= 22 - (4 + 10)

= 8.

94

$$\text{Free float late} = \frac{\text{latest date for}}{\text{succeeding event}} - \frac{\text{latest date for}}{\text{preceding event}} + \text{duration}$$
$$= 22 - (8 + 10)$$
$$= 4.$$

17 − 8 + 9

Resource levelling

Most contracts are won on the basis that the firm submitting the lowest tender gets the job. The examples so far have been concerned with establishing a logical sequence between a number of activities and then calculating the duration of the project. Times allowed for activities would be based on historical records whenever possible. If the firm has no previous experience of an activity, its duration would be estimated by normal estimating techniques. So far no consideration has been given to achieving a high utilisation of resources. After the first network has been prepared, it is necessary to examine the resource requirements of that particular programme. It is highly likely that the firm will place a limit on the resources which can be spared for a project. If the programme as prepared requires one tower crane for 12 months and a second one for a duration of 3 months in the middle of the period, some way must be found to manage with one crane.

Example 5.12

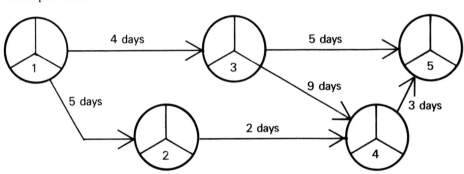

Table 5.7

Resources for Example 5.12

Activity	Duration	Men required	Plant required
1 — 2	5	10	A
1 — 3	4	20	B
2 — 4	2	4	C
3 — 4	9	2	C
3 — 5	5	5	B
4 — 5	3	4	C

Table 5. 8

Activities schedule for Example 5.12

Activity	Duration	Start Earliest	Start Latest	Finish Earliest	Finish Latest	Total Float
1 — 2	5	0	6	5	11	6
1 — 3	4	0	0	4	4	0
2 — 4	2	5	11	7	13	6
3 — 4	9	4	4	13	13	0
3 — 5	5	4	11	9	16	7
4 — 5	3	13	13	16	16	0

95

The resources available are:
1. A maximum of 20 men.
2. Plant: 1 A type, 1 B type, 1 C type.

The next step is to plot the information on a Gantt Chart.

Table 5.9

Example 5.12 plotted on a Gantt Chart

Days	1	2	3	4	5	6	7	8	9	10	11	12	13	14	15	16	17	18	19
1 — 2 (10A)		1				2													
1 — 3 (20B)		1			3														
2 — 4 (4C)						2		4											
3 — 4 (2C)						3							4						
3 — 5 (5B)						3				5									
4 — 5 (4C)														4			5		

Men / Resources	1	2	3	4	5	6	7	8	9	10	11	12	13	14	15	16
Men	30	30	30	30	17	11	11	7	7	2	2	2	2	4	4	4
A	1	1	1	1												
B	1	1	1	1	1	1	1	1								
C					1	1	2	2	1	1	1	1	1	1	1	1

Table 5.9 shows all activities plotted at their earliest start time. The activities are shown with rectangular outline whilst the float associated with each activity is shown as a straight line.

Demand for men

The demand for men in the first four days exceeds the supply, whilst the demand for resource 'C' exceeds the supply on two days, namely days 6 and 7. If it is decided to concentrate on levelling the manpower first, it should be noted that activity 1 - 3 is critical and therefore should not be moved unless there is no alternative.

Table 5.10

Resources smoothed

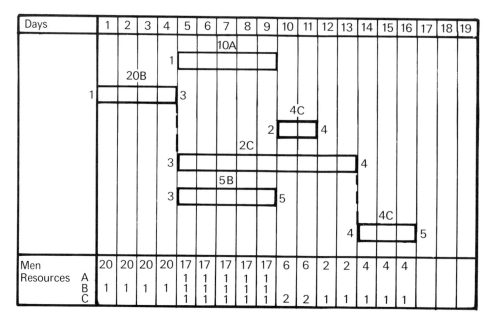

Men / Resources	1	2	3	4	5	6	7	8	9	10	11	12	13	14	15	16
Men	20	20	20	20	17	17	17	17	17	6	6	2	2	4	4	4
A					1	1	1	1	1							
B	1	1	1	1	1	1	1	1	1							
C					1	1	1	1	1	2	2	1	1	1	1	1

Activity 1 - 2 possesses float and should be moved to the right four days. This considerably improves the demand for men. Days 6 and 7 require 21 men, otherwise the manpower problem is solved. By moving activity 2 - 4 to the right to occupy days 10 and 11, the demand for men never exceeds 20 (see Table 5.10).

Demand for plant

The demand for the C resource on days 10 and 11 exceeds the supply. None of the activities using the C resource can be moved to the left. Activity 2 - 4 must follow activity 1 - 2 and activity 3 - 4 must follow after 1 - 3. An activity using a C resource must be moved to the right. Activity 2 - 4 is now moved to days 14 and 15 and activity 3 - 4 to days 16, 17 and 18. Once brought on site, there is a continuous demand for each of the three resources.

Table 5.11

Demand for 'C' levelled

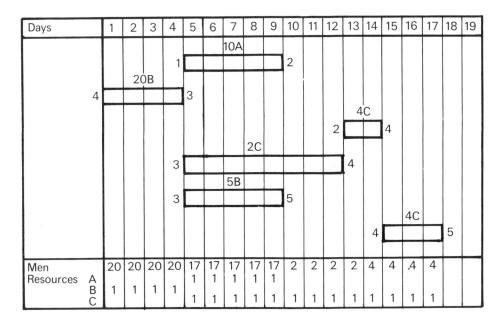

Days	1	2	3	4	5	6	7	8	9	10	11	12	13	14	15	16	17	18	19
Men Resources A	20	20	20	20	17	17	17	17	17	2	2	2	2	4	4	4	4		
B	1	1	1	1	1	1	1	1	1										
C					1	1	1	1	1	1	1	1	1	1	1	1	1	1	

To achieve the required objective, the programme has been lengthened by two days. The demand for plant has been 'levelled', whilst the demand for men has been 'smoothed'.

Crash costing

Example 5.12 is concerned with rationalising the demand for resources in order to minimise cost, a very common objective. However, there are times when it is financially sound for the client to pay the contractor extra money to finish earlier than he would do otherwise.

The contractor preparing a network may find that his first attempt overshoots the time available and methods and target times must be reviewed. The critical path method provides a useful technique for achieving the required shortening of the contract at the least extra cost. Only those activities which are on the critical path should be considered. To shorten non-critical activities would simply be buying float. It may not be possible to shorten all critical activities, therefore a list should be made of the cost for each day saved. Normally, those with least extra cost would be used, in rank order, until the required reduction was secured.

Example 5.13

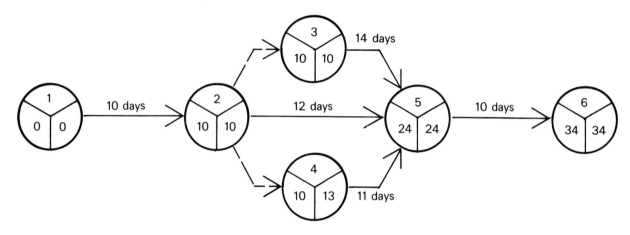

The required contraction is best viewed as a series of steps so that a constant check is made on the location of the critical path. With each day saved on the critical path, the float on some non-critical item is shortened by the same amount and ultimately such non-critical activities become critical.

Table 5.12

Activity	Normal time days	Compression time available	Normal Cost £	'Crash' Cost £	Cost per day £ saved
1 — 2	10	0	1,000	—	—
2 — 5	12	1	800	850	50
3 — 5	14	3	1,600	1,720	40
4 — 5	11	4	1,100	1,200	25
5 — 6	10	2	1,000	1,090	45
		Total	5,500		

Table 5.13   **Activities list for Example 5.13**

Activity	Duration	Start Earliest	Start Latest	Finish Earliest	Finish Latest	Total Float
1 — 2	10	0	0	10	10	0
2 — 5	12	10	12	22	24	2
3 — 5	14	10	10	24	24	0
4 — 5	11	10	13	21	24	3
5 — 6	10	24	24	34	34	0

Using the Compression Table 5.14

Before step 1 there are three activities on the critical path, but one of these has no compression time available, namely (activity 1 - 2). In considering the other two, the cost is £40 with both of them, so in this respect they are equally acceptable. However, as 3 - 5 is the earliest in the contract, this is to be preferred, since effecting the 'saving' early would relieve anxiety.

Table 5.14 Step by step compression

Activity Number	Cost per day	Before Step 1 Compression Time Available	Before Step 1 Float	After Step 1 Compression Time Remaining	After Step 1 Float	After Step 2 Compression Time Remaining	After Step 2 Float	After Step 3 Compression Time Remaining	After Step 3 Float	After Step 4 Compression Time Remaining	After Step 4 Float	After Step 5 Compression Time Remaining	After Step 5 Float
1—2	—	0	0	0	0	0	0	0	0	0	0	0	0
2—5	50	1	2	1	1	1	1	1	1	1	1	0	0
3—5	40	3	0	2	0	1	0	1	0	1	0	0	0
4—5	25	4	3	4	2	4	1	4	2	4	2	4	2
5—6	40	2	0	2	0	2	0	1	0	0	0	0	0
Cost		£5500		£5540		£5580		£5625		£5670		£5760	
Duration		34 days		33 days		32 days		31 days		30 days		29 days	
Activity Shortened		—		3 — 5		3 — 5		5—6		5—6		2—5 3—5	

Step 1

Shorten activity 3 - 5. Note that this action reduces the float by one day each on activities 2 - 5 and 4 - 5.

Step 2

Shorten activity 3 - 5. This action causes activity 2 - 5 to become critical. There is a double critical path, since activities 2 - 5 and 3 - 5 now have the same duration of 12 days. Any further shortening in this part of the network must include both these activities.

Step 3

A choice is available. Either shorten activity 5 - 6 alone or both activities 2 - 5 and 3 - 5.

	(a) £		(b) £
Cost: 5 - 6	45	2 - 5	50
		3 - 5	40
	45		90

Shorten activity 5 - 6, which is the cheapest.

Step 4

Shorten activity 5 - 6. There is no further compression time available to activity 5 - 6.

Step 5

Shorten activities 2 - 5 and 3 - 5 to save one further day. Note that all three activities in the middle of the diagram are now critical, with eleven days duration each. This means that all five activities in the network are now critical.

Comparison between direct and indirect costs

If it is assumed that apart from the direct costs of each activity, there are indirect costs of £50 per day, this would amount to £1700 for the whole job based on the normal contract time of thirty four days. If the contract exceeds its normal duration, it is assumed that these costs continue to accrue. Figure 5.11 shows the direct cost if the contract is shortened whilst Figure 5.12 illustrates the indirect costs. Figure 5.13 is the summation of Figures 5.11 and 5.12.

Figure 5.11   **Direct costs**

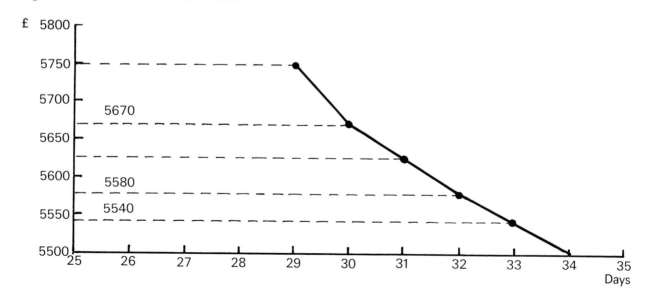

Figure 5.12   **Indirect costs**

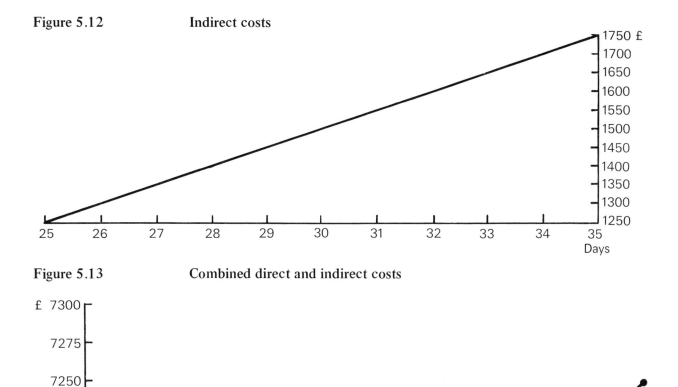

Figure 5.13   **Combined direct and indirect costs**

Table 5.15

Costs	Days							
	29	30	31	32	33	34	35	36
Direct £	5760	5670	5625	5580	5540	5500	5500	5500
Indirect £	1450	1500	1550	1600	1650	1700	1750	1800
Total	7210	7170	7175	7180	7190	7200	7250	7300

Figure 5.13 and Table 5.15 show that the most favourable contract duration is 30 days.

SUBNETWORKS

A subnetwork is a magnification of part of a network. A single activity shown on an overall programme network, when examined in detail, may be very complex and in fact made up of a number of small interrelated activities. This can be illustrated in a network diagram which forms a subnetwork to the main one. The Ewood drainage example is illustrated by means of a Gantt Chart but operation number 7, laying the sewer in Dayspring Lane, could usefully be the subject for a subnetwork or other form of detailed scrutiny. It is planned to do this work in one week and it contains approximately 20 separate activities.

Table 5.16   **Activities List**

1. Communicate with the statutory bodies, Water, Gas, Electricity, Telephone, Foul sewers, Storm water, to find if they have buried services in the area which it is proposed to dig.

2. Make application to the local authority for a temporary street closure and discuss the re-routing of traffic, the provision of barriers, temporary signs and temporary lighting.

3. Make signs.

4. Fix signs but cover up.

5. Unveil signs and position barriers.

6. Break into the road.

7. Redirect electricity cable.

8. Dig trench.

9. Lay loose fill bed.

10. Lay sewer pipes.

11. Test sewer.

12. Complete loose fill surround.

13. Back fill trench.

14. Re-test sewer.

15. Make temporary road surface.

16. Remove barriers.

17. Remove temporary signs.

The agreement with the local authority and police would probably be that the temporary route should become operative from 0500 hours on Monday and revert to the original route at the same time one week later. Note that activities 1, 2, 3 and 4 can be carried out, if necessary, weeks before the work actually starts. Activities 5, 16 and 17 must be carried out at a precise time. Not all activity durations are shown in hours on the network.

The one week duration of this part of the total project relates to the commencement of activity 5 on the list to activity 17. That is, from 'unveil signs' to 'remove barriers and signs'. In hours, this is 301 − 245 = 56. (See Figure 5.14.) This would mean obtaining permission to work overtime from the local 'overtime committee' and possibly working a ten hour day for six days.

When an event appears in two places, in this instance event number nine, it should have a thicker ring than the others. It is referred to as 'interface'.

TIME/PRODUCTION GRAPHS

The planned progress and actual progress of individual trades can be illustrated graphically in the manner shown in Figure 5.15. This method of control is very suitable for those trades which are easily identified as the key trades in a project.

Figure 5.14   Sub-network for work in Dayspring Lane

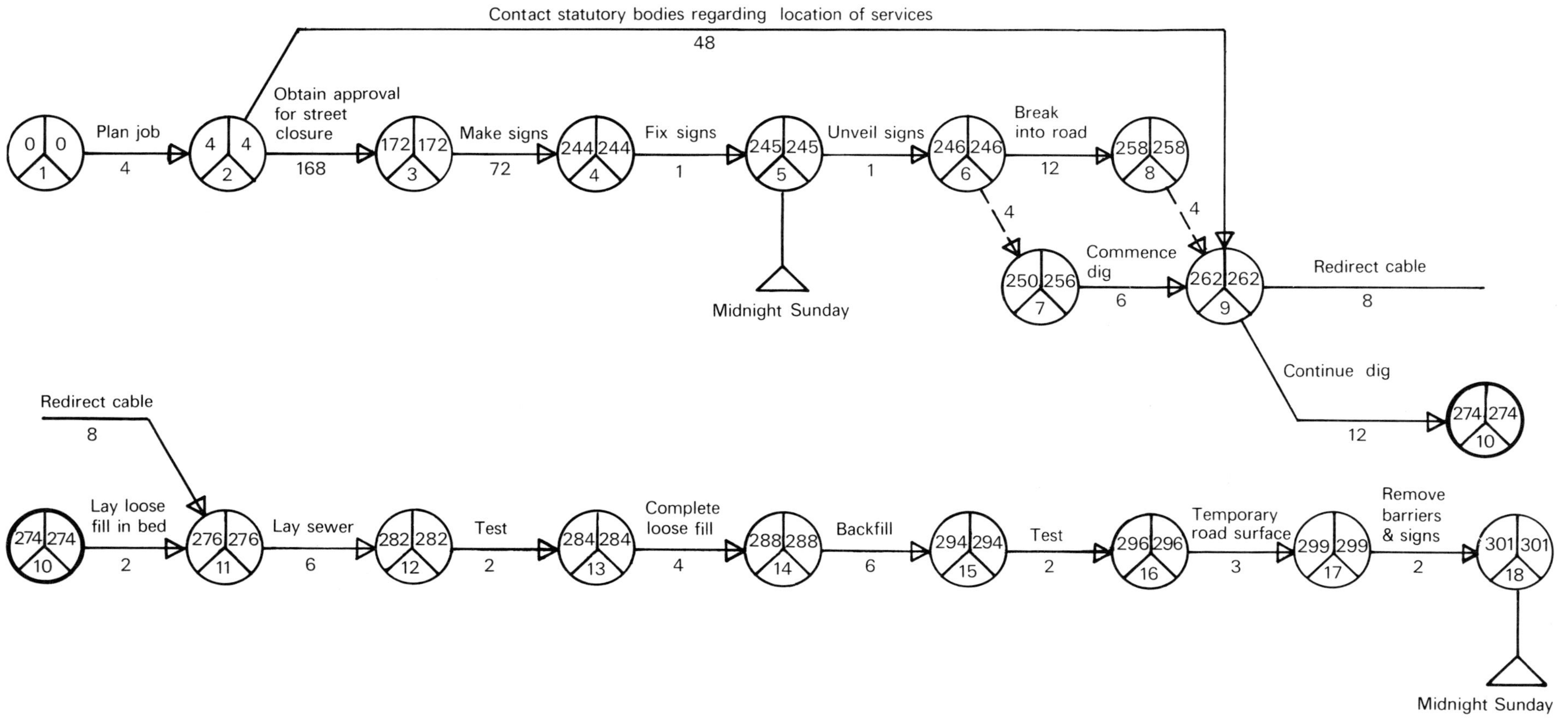

For example, assume that it is planned to do the painting and decorating of 35 houses in 12 weeks with a labour force of 10 men. This is shown in Figure 5.15 by line OF. By the end of week 4 only 7 houses have been completed.

The production rate is therefore:

$$\frac{7 \text{ houses}}{4 \text{ weeks}} = 1\frac{3}{4} \text{ houses per week.}$$

The planned rate of production was:

$$\frac{35 \text{ houses}}{12 \text{ weeks}} = 2\frac{11}{12} \text{ houses per week.}_{\mathsf{r}}$$

If no corrective action is taken, the painting and decorating will be completed in:

$$\frac{35 \text{ houses}}{1\frac{3}{4} \text{ houses per week}} = 20 \text{ weeks.}$$

To have the job completed by the original target date of week 12 will require an adjustment to the labour strength.

Figure 5.15   **Time/production graph**

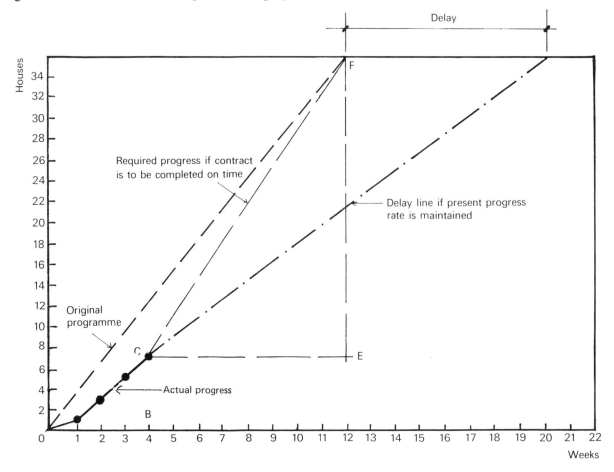

Calculating the adjustment to the production rate.

Remaining production	= 35 – 7 houses
	= 28 houses.
Remaining time	= 12 – 4 weeks
	= 8 weeks.

$$\text{Required production rate} = \frac{28 \text{ houses}}{8 \text{ weeks}}$$

$$= 3\tfrac{1}{2} \text{ houses per week.}$$

To calculate the adjustment to the labour strength

$$= \frac{\text{required production rate}}{\text{present production rate}} \times \frac{\text{current labour strength}}{1}$$

$$= \frac{3\tfrac{1}{2}}{1\tfrac{3}{4}} \times \frac{10}{1}$$

$$= 20 \text{ men.}$$

An increase of 10 men is required.

Reverse use of time/production graphs

If sales on a development site are not reaching the anticipated level, it may become necessary to slow down production. Empty houses are un-profitable from all points of view. In order to control the transition to a reduced level of output, it is necessary to slow down the finishing trades first. The graph (Figure 5.16) relates to the painter and decorator trade. It can be seen from the graph that the handover rate was consistent at around 13 completions per month. Handovers commenced at the beginning of month 10 and were planned to be completed by month 19.

$$\frac{120 \text{ houses}}{39 \text{ weeks}} = 3\tfrac{1}{13} \text{ or 3 houses per week.}$$

By month fifteen, sixty-five houses had been completed. A production rate of $\frac{65 \text{ houses}}{22 \text{ weeks}} = 3$ houses per week.

By month fifteen, it is decided to slow the rate of completions down so that the last houses are handed over by the twenty-fourth month, see Figure 5.16.

Figure 5.16   **Reverse use of time/production graph**

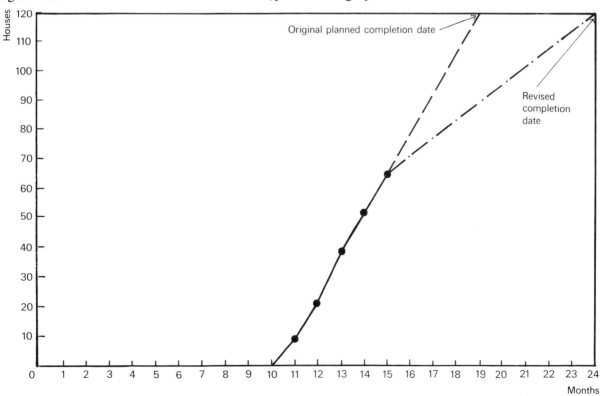

105

Houses remaining $= 120 - 65$
$= 55.$

Time remaining $= 24\ \text{months} - 15$
$= 9\ \text{months}$
$= 39\ \text{weeks}.$

Required production rate $= \dfrac{55\ \text{houses}}{39\ \text{weeks}}$

$= 1\frac{16}{39}\ \text{houses per week}.$

Assume that the current labour force is 16 men.

New labour force $= \dfrac{\text{required production rate}}{\text{original production rate}} \times \dfrac{\text{current labour force}}{1}$

$= \dfrac{1\frac{16}{39}}{3} \times \dfrac{16}{1}$

$= 6.56$

$= 7\ \text{men}.$

Recording progress on a large housing site

On a small site the site manager would be familiar with the details of the progress of every house or flat. On a large site it is not so easy and the site manager will require a proper record. When making his routine daily or twice daily inspection, it is advisable to carry a clip board with control sheet (see Table 5.17).

Table 5.17

Recording progress

Contract: **Recorded by:**

Date:

Stages completed this week

Plot numbers
1 2 3 4 5 6 7 8 9 10 11 12 13 14 15 16 17 18

Concrete foundations	x x x x x x x x x x √ √ √ √ √ √
Brickwork to dpc	x x x x x x x x √ √ √ √ √ √ √
Service entries	x x x x x x x √ √ √ √ √ √ √ √
Hardcore fill	x x x x x x x √ √ √ √ √ √ √
Sub floor ducts	x x x x x x √ √ √ √ √ √ √
Oversite concrete	x x x x x √ √ √ √ √ √
Brickwork 1st lift	x x x x √ √ √ √ √ √ √ √
Brickwork 2nd lift	
Joists	
Brickwork 3rd lift	
Brickwork 4th lift	
Roof timbers	
Brickwork cutting to roof	
Roof tiling	
Plumbing to roof	

Note: x represents stages completed previously.
√ represents stages completed this week.

Once the job establishes a rhythm, the manager will be able to predict with a high degree of accuracy the amount of work which each group of sub-contractors or directly employed men will complete during the next two weeks. This is very useful when considering the call forward of materials or the last date for receiving sales information from clients, ie, colour of internal decorations, etc. It is also useful as a guide to what sums the sub-contractors will be including in their weekly or monthly payments, and thus what work requires snagging before payments are certified.

LINE OF BALANCE

Line of Balance is a production planning technique which is suitable only for repetitive construction. One of its positive advantages over many other techniques is that both overall and detailed aspects of planning and control are displayed on the one chart. On the debit side a degree of control is pre-supposed over all labour gangs, which seldom exists in practice. No account appears to be contemplated that the variable factors such as adverse weather, acute shortage of labour in certain trades, materials delays, do not affect all trades equally.

All activities are deemed to be critical. Activities are considered as a single chain sequence.

Figure 5.17

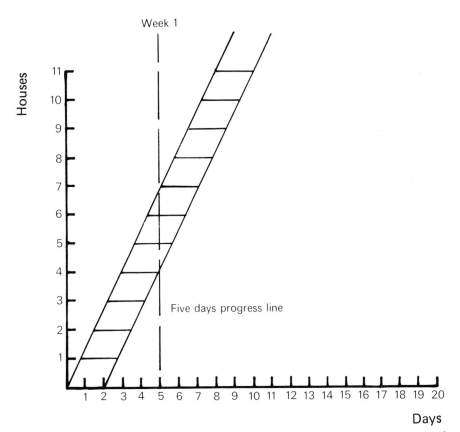

In a factory assembly line, at each point in the assembly a stock of spare components is maintained to ease out problems of different rates of assembly. In construction work there cannot be any spare floors or walls, so time buffers are inserted between each trade, which has the same effect of creating a buffer of work between them.

Gang sizes must be adjusted so that the weekly production of each gang expressed in terms of units per week is as near as possible identical.

Some ideas about 'balance'

Assuming that the painter and decorator are the last trade involved (in some cases it would be the floor tiler), this site will be in progress for the length of time that it takes to complete the first house ready for painting, plus sixty-three weeks. To prepare the first house up to painting stage would take ten to fifteen weeks. If this (say seventy-five weeks) is deemed too long, then the gang sizes must be increased pro rata. Doubling the gang sizes would reduce the contract period by the reduced amount of time, if any, that the first house is ready for painting plus thirty-one and a half weeks. Because doubling the gang sizes will not necessarily increase the number of men working in one house, it may still take 12 to 15 weeks to have the first house ready for the painter and decorator. Next, the planner must know at what rate per week or month the houses must be ready to hand over to meet the sales programme.

Table 5.18

Trade	Hours for one house	Hours for fifty houses	Gang size	Gang Hours on site	Gang weeks
Bricklayer	250	12,500	6	2,083	52
Carpenter and Joiner	320	16,000	8	2,000	50
Plasterer	150	7,500	3	2,500	63
Plumber and Heater	100	5,000	2	2,500	63
Painter and Decorator	100	5,000	2	2,500	63

Application of Line of Balance Theory

The example (Tables 5.19 and 5.20 and Figures 5.18 and 5.19) relates to seven blocks of maisonnettes each containing four dwellings, a total of twenty-eight dwellings. The work is of brick and block construction with cast insitu concrete floors at first floor level. The programme relates to the work above the horizontal dpc. A number of trades are not programmed because they could easily be controlled within the time range of activities which are programmed. For example, the cast insitu first floors are not programmed. When the bricklayers raise a block to first floor level, they move over on to another block whilst the first floor is prepared and the block vacated. This includes formworker, electrician for conduit, steel-fixer, and finally concretor. These four activities can be carried out within the time for bricklaying without causing that trade any delay, and also without delaying the following trade of roof carpenter.

Some activities are of very short duration and tend to unbalance the programme, for example the durations of the floor insulation and screeding, wall tiling, and artex ceiling operations. In minimising the amount of time buffer which develops around these activities, success depends upon the number of visits that these specialists are prepared to make to the site. Insulator and floor screeder make separate visits to each block, whilst wall tiler and artex ceiling contractor complete two blocks each visit.

Table 5.19 Progress check sheet

Operation	Man hours per dwelling A	Theoretical gang size at three dwellings per week B	Men per dwelling C	Actual gang size D	Actual rate per week E	Time in days for one dwelling F	Time between starting work on first and last dwellings G
Brick and block work	100	7.5	6	6	2.4	2.08	56
Roof carpentry	16	1.6	2	2	5.0	1.0	27
Roof tiling	24	1.8	2	2	3.0	1.5	45
Plumbing 1st Fix	40	3.0	1	3	3.0	5.0	45
Joiner 1st Fix	60	4.5	2	4	2.7	3.75	50
Heating engineering carcase	20	1.5	2	2	4.0	1.25	34
Electrical carcase	24	1.8	2	2	3.0	1.5	45
Floor insulation and screeding	10	0.75	2	2	8.0	0.5	18
Dry lining	60	4.5	2	4	2.7	3.75	50
Joiner 2nd Fix	80	6.0	2	6	3.0	5.0	45
Electrician 2nd fix	20	1.5	2	2	4.0	1.25	34
Heating 2nd fix	40	3.0	1	3	3.0	5.0	45
Plumber 2nd fix	45	3.4	2	4	2.7	2.8	50
Wall tiling	6	0.5	1	1	6.7	0.75	20
Artex ceiling	8	0.75	1	1	5.0	1.0	27
Decoration	60	4.5	2	4	2.7	3.75	50
Cleaning and floor tiling	8	0.75	1	1	5.0	1.0	27
Snagging and handing over	8	0.75	1	1	5.0	1.0	27

It should be noted that the first three trades, brick and blockwork, carpenter and roof tiler, treat one block as a whole and not four separate dwellings. Whilst a block is in course of erection, scaffolded, etc, the bricklayer dominates activity on the whole block. He does not act to complete dwellings individually.

Figure 5.18

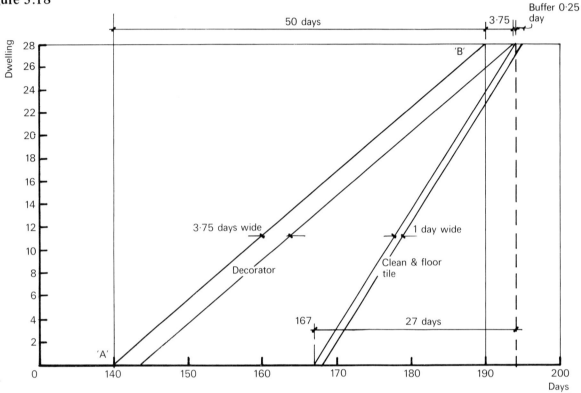

Figure 5.19　　　　**Progress to day 180**

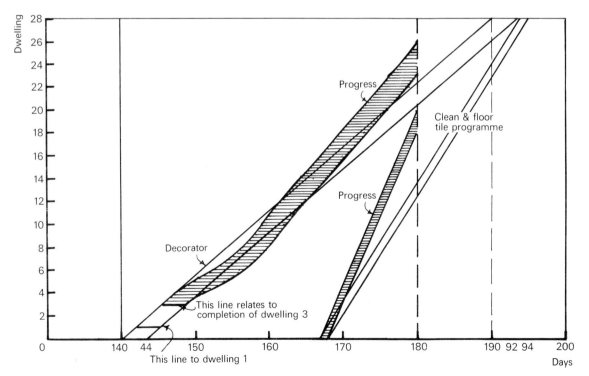

Table 5.20		Particulars related to one block		
Gang size	*Trade*	*Details*		*Time (days)*
6	Bricklayers	4 scaffold lifts		2 per lift
2	Scaffolders	3 scaffold lifts		1 per lift
4	Formworkers	Suspended floor		2 per floor
1	Electrician	Conduit to suspended floor		½ per floor
2	Steelfixers	Suspended floor		1 per floor
5	Concretors	Pumped concrete to floor		½ per floor

Calculations

Example: brick and blocklayer

There are a total of 168 000 bricks in the seven buildings. This includes blockwork converted to brick equivalent.

The estimated performance rate is 60 bricks per hour.

$$\text{Man hours for 28 dwellings} = \frac{168\ 000}{60}$$

$$= 2800.$$

$$\text{Man hours for 3 dwellings} = 300.$$

$$\text{To accomplish this in 40 hours requires } \frac{300}{40} \text{ men}$$

(Column B)

$$= 7.5 \text{ men.}$$

However, the sub-contractor has a gang of 6 men. These men work together as one gang; there is ample wall length on a block for their safe and efficient operation.

The actual rate per week is a relationship between the theoretical gang size and the actual gang size, and is found as follows:

$$\text{Actual rate} = \frac{\text{actual gang size}}{\text{theoretical gang size}} \times \frac{\text{planned rate}}{1}$$

$$= \frac{6}{7.5} \times \frac{3}{1}$$

$$= 2.4 \quad \text{(Column E)}$$

The time in days for one dwelling is found by dividing the hours in one dwelling by the men per dwelling and converting this hourly total into days.

$$\text{Time in days} = \frac{100 \text{ hours}}{6 \text{ men}} \div 8 \text{ hours}$$

$$= 16.7 \div 8 \text{ hours}$$

$$= 2.08 \text{ days} \quad \text{(Column F)}$$

Column G. The time which elapses between the start of an activity on the first dwelling and the start of the same activity on the last is found as follows:

$$\frac{(\text{total number of dwellings} - \text{the last dwelling}) \times 5 \text{ days}}{\text{actual rate per week}}$$

$$= \frac{(28 - 1) \times 5 \text{ days}}{2.4 \text{ days}}$$

$$= \frac{135}{2.4}$$

$$= 56 \text{ days.}$$

The production rate of roof carpenter is more than twice that of brick and blockworker. When the 'following trade' is proceeding more quickly, the time buffer is placed at the top. In this case, it is deemed necessary to ask the roof carpenter to carry out the roofs for the first four blocks, then leave the site and return later to complete the remaining three blocks. The position of the five-day time buffer should be noted. The carpenter carcase is proceeding at a slower rate than the plumber carcase. In this case the time buffer is placed at the bottom. Note the four visits to site of wall tiler and artex ceiling worker. Had they been planned to do all their work in one stay on site, the time buffer between them both and plumber second fix would have been in the region of thirty working days.

Plotting the chart

Figure 5.18 shows how the chart is plotted by illustrating the plotting of activities 'decoration' and 'clean and floor tile'.

The plotting figures are obtained from Table 5.19, with the exception of the point 'A'. This point is obtained by applying an agreed time buffer between the start of 'wall tiler and artex ceiling' and the start of 'decorator'. Having established this point, day 140, reference is made to column G of Table 5.19. This states that 50 days elapse between the start on dwelling one, 'decorate', and the start of 'decorate' to dwelling twenty-eight. By referring to column F and noting that the gang time for decorating one dwelling is 3.75 days a line can be drawn parallel to AB and 3.75 days distance from it.

Clean and floor tile proceeds at a faster rate than decorator (see Table 5.19, column E). The time buffer between the two is inserted at the top, in this case 0.25 days. This locates day 194; the time between starting cleaning dwelling one and cleaning dwelling twenty-eight is 27.0 days (column G).

This is 194 − 27 = day 167, and locates the bottom of the line for 'clean and floor tile'. The chart can be completed by drawing the line 0.5 of a day from this to display the finishing time for each 'clean and floor tile' activity.

Marking up progress

Figure 5.19 shows the progress marked up to day 180. According to the original plan, work should have been taking place on dwelling numbers 20, 21 and 22. In fact, they are completed and work now proceeds on dwellings 24, 25 and 26. This trade is ahead of schedule. Note should be made of the period, day 150 to 164. This trade was getting behind schedule.

Clean and floor tiler programme

This work is well ahead of schedule. According to the original programme dwelling 12 should just have been tiled, but, in fact, work is now proceeding on dwelling number 19.

If these trades represented concreting foundations, building and plastering brick walls, the practical effects of this would be as follows:

1. Concreting foundations would be ahead of schedule.

2. The buffer between bricklaying and plastering would be reduced.

Summary of Line of Balance

1. There are not many main contractors who directly employ more than two or three trades on housing work. Housing is probably the largest supply of repetitive work.

Figure 5.20

Procedure diagram planning

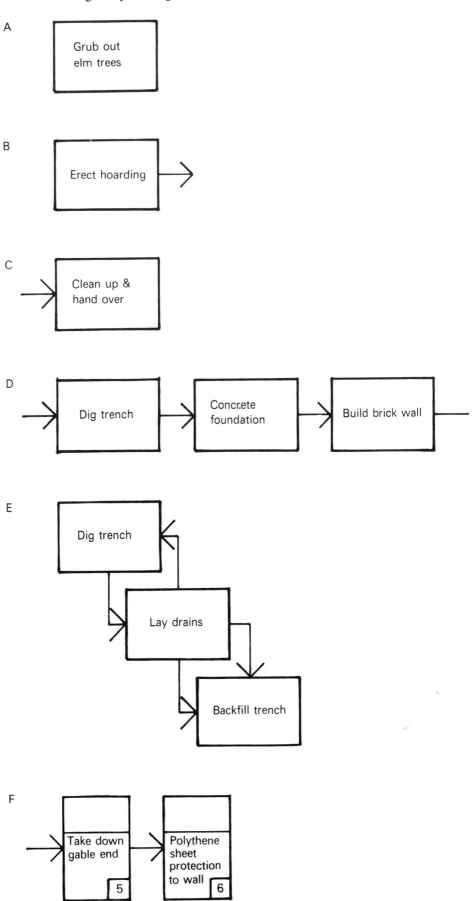

2. When sub-contractors are employed, it is not easy for the main contractor to state precisely how many men are employed by each sub-contractor. The sub-contractor may organise his workmen in gangs of four. Why should he change to satisfy the needs of one site?

3. On one recent housing contract, it was observed that seven trades commenced work within a few days of each other once the timber frames had been erected. This is difficult to display on a line of balance diagram, and even more so if some trades get behind and some in front of programme. Advocates would suggest the use of a colour code for mark up; this may be useful until it is necessary to produce a revised programme, which, in many offices, would be dye-line printed in monochrome.

SEQUENCE STUDIES

Some site managers prefer to use a Gantt Chart for general control and this is then supported by a study in detail for any interrelated repetitive trades which are present. In many buildings, the careful control of the in-situ concrete frame is of paramount importance to the successful completion of a contract. Success often depends on maximising the use of framework. In the construction of the seven blocks of maisonnettes the cast in-situ concrete first floor is supported on the inner leaf of the cavity wall and upon certain load-bearing partitions. Steel conduit for the lighting system to the ground floor dwellings is cast into the concrete.

Table 5.20 shows the trades that were involved in building the unroofed shells.

A piece of squared paper is useful for the plotting. (Table 5.21) The following constraints should be noted:

1. Brickwork must be not less than 24 hours old before any scaffolding depends upon it.

2. Electrician precedes steelfixer.

3. Formwork to suspended floor should not be removed before seven days.

4. The main priority must be to achieve 100% utilisation of the bricklayer's time.

5. A second line priority is to operate on the least number of blocks possible in order to minimise the amount, and thus the cost of scaffolding to the job.

The reader should study the chart carefully. Note that no two groups of workers operate on the same block at the same time.

The sequence as depicted by the chart is:

1. Build first lift brickwork.

2. Scaffold for second lift brickwork.

3. Build second lift brickwork.

4. Scaffold for third lift brickwork.

5. Erect formwork for suspended concrete floor.

6. Position steel conduit for lighting system.

7. Position steel reinforcement for suspended floor.

8. Place concrete.

114

Table 5.21

Sequence Study - 8 week period

	WEEK 1	WEEK 2	WEEK 3	WEEK 4
Bricklayer				
1st Lift	A B C D E			
2nd Lift			A B C D	E
3rd Lift				
4th Lift				
Scaffolder				
1st Lift		A B C D E		
2nd Lift				A B
3rd Lift				
Formworker				
Electrician				
Steelfixer				
Concretor, floors				

	WEEK 5	WEEK 6	WEEK 7	WEEK 8
Bricklayer				
1st Lift	F G			
2nd Lift		F G		
3rd Lift			A B C D	E
4th Lift				A
Scaffolder				
1st Lift		F G		
2nd Lift	C D E		F G	
3rd Lift			A B C	D E
Formworker	A B C D	E		F G
Electrician				
Steelfixer		A B	C D E	F
Concretor, floors				

115

9. Build third lift brickwork.

10. Scaffold for fourth lift brickwork.

11. Build fourth lift brickwork.

Items which are not included but form part of the work are, strip out and clean formwork and take down scaffolding. Scaffolding would of course be adapted for the roofing operation before being removed.

THE ADVANCING FRONT

Another type of display which is used by some housing developers is called the advancing front. This necessitates identifying, by a number, all the different activities in the house. These should be numbered in the sequence in which trades start. The separate houses or terraces of houses should be identified by a letter of the alphabet and in alphabetical order (see Table 5.22).

When the progress check is carried out, operations are ticked off if completed. A piece of perspex can be cut to the planned profile of completions and imposed on the diagram to check on any variations from plan, hence the name the 'advancing front'.

PRECEDENCE DIAGRAM PLANNING

It is claimed by some that critical path diagrams are not easy to read and that they are an unfair imposition on the busy site supervisor. Precedence diagram planning is said to contain all the logic aspect of the critical path method, but is in fact easier to read. Activities are depicted by a box (see Figure 5.20 A). If the activity is a starter or commencing activity, it will be shown as in Figure 5.20 B. Terminal activities are shown in Figure 5.20 C. If they are ordinary single activities, they are shown as in Figure 5.20 D.

When it is unnecessary to dig the whole of a trench before drainlaying commences and backfilling can commence before all the drains are laid, there are overlap boxes (see Figure 5.20 E). Each box should have an identification number (see Figure 5.20 F). When two or more activities merge into one activity, this activity is said to be a merge box (see Figure 5.21 A). It should be noted that the boxes are numbered from left to right as with event points in critical path planning. The numbers also denote the order in which the activities will be shown on the cascade bar chart. On completion of an activity, it is sometimes possible for two or more activities to start (see Figure 5.21 B); this depicts a burst box.

The system of box numbering shown in Figure 5.20 C should be noted. However, if there is any doubt when a burst box is reached, the longest timewise chain to the next merge box should be numbered out first.

Time

The diagrams so far have been concerned with depicting logic only. It is now necessary to introduce a complete system to cover activity and event times, activity duration and float (see Figure 5.22 A).

Note the example shown in Figure 5.22 B. If this logic were drawn as a critical path diagram, two dummies would be required to preserve the unique numbering system. Dummies are not necessary in precedence diagram planning. Figure 5.22 C shows how the amount of overlap is denoted.

116

The calculations for precedence diagrams are made in exactly the same way as for critical path.

Readers should note carefully the numbering of the boxes in Figure 5.23. 'Demolish building' is box number 2 in preference to 'site scrape' because it has a longer time duration. Similarly, 'excavate service trenches within building' and 'lay services within building' together take ten days as against the three days of 'bore piles' and 'pour piles'. The longest timewise chain is numbered first.

Table 5.22

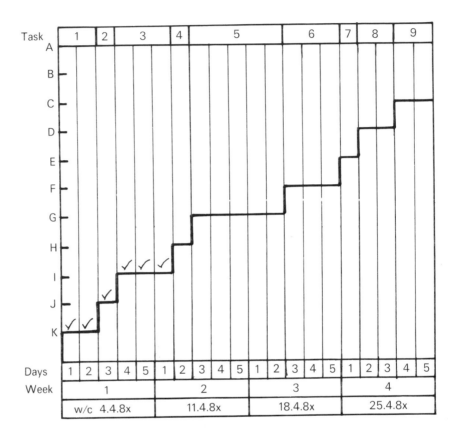

Cascade bar chart

Figure 5.24 is a cascade bar chart derived from the precedence diagram Figure 5.23. Correct numbering of the boxes ensures a proper cascade of the critical activities. Activities numbered 8, 9 and 10 share 5 days of total float. Activities 3, 12 and 14 each have free float.

117

Figure 5.21

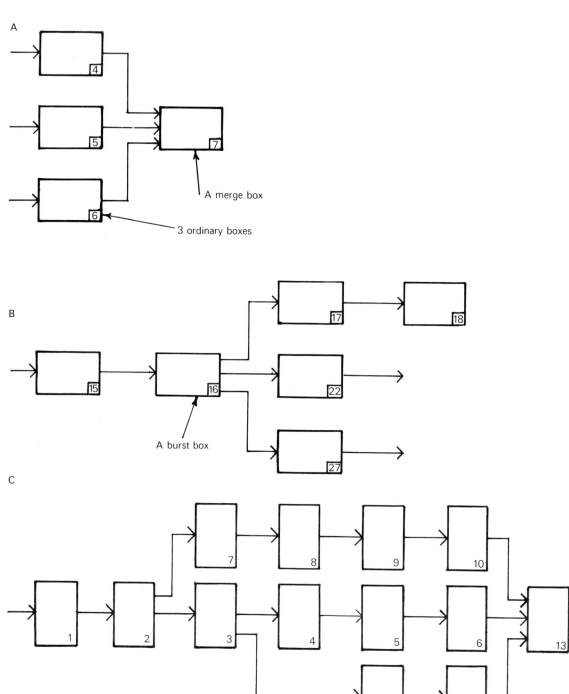

A

A merge box

3 ordinary boxes

B

A burst box

C

118

Figure 5.22

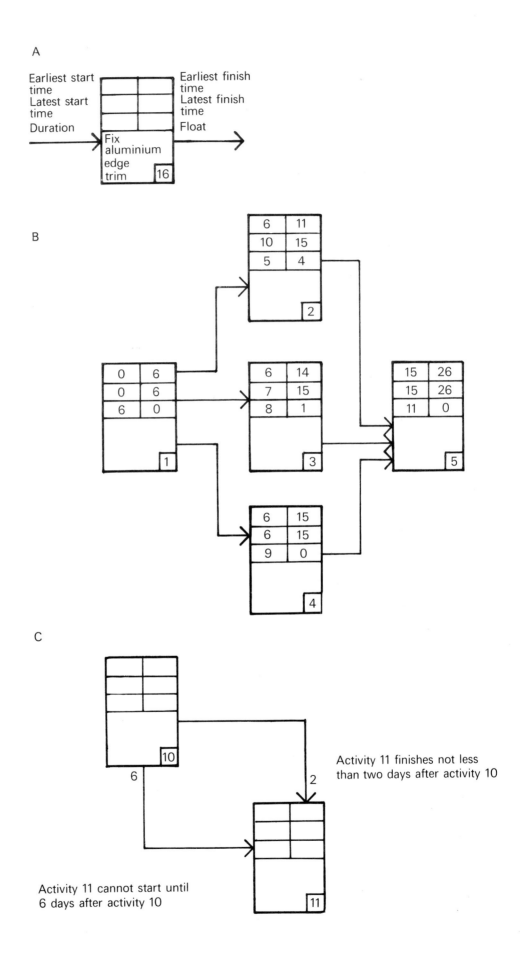

Figure 5.23

Precedence diagram

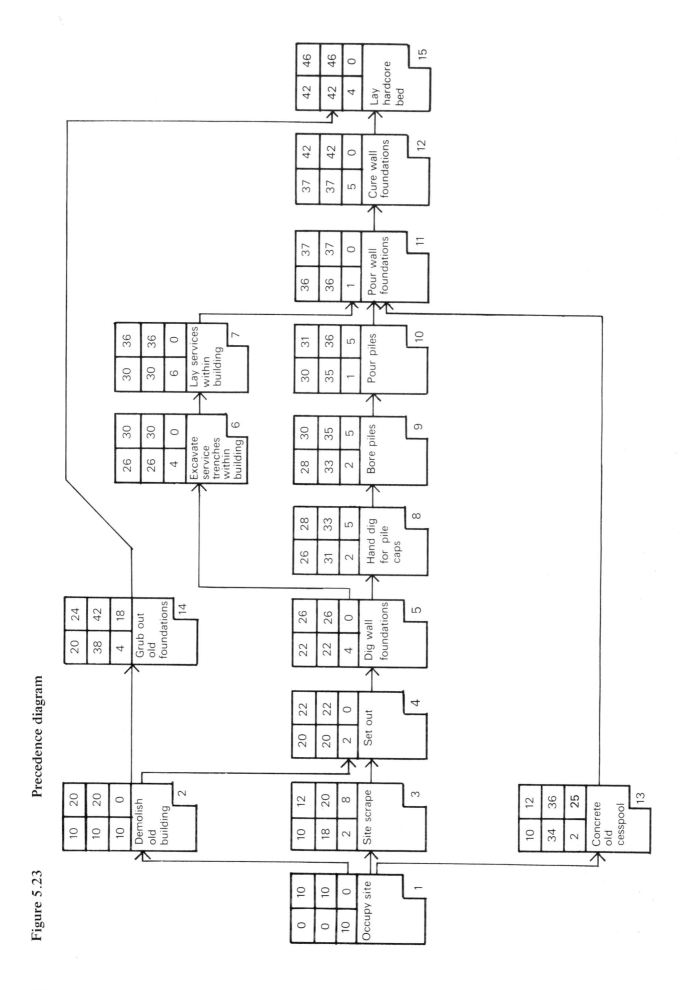

Figure 5.24   Cascade bar chart

		1	2	3	4	5	6	7	8	9	10	11	12	13	14	15	16	17	18	19	20	21	22	23	24	25	26	27	28	29	30	31	32	33	34	35	36	37	38	39	40	41	42	43	44	45	46
1	Occupy site																																														
2	Demolish old building																																														
3	Site scrape																																														
4	Set out																																														
5	Dig wall foundations																																														
6	Excavate service trenches																																														
7	Lay services within building																																														
8	Hand dig for pile caps																																														
9	Bore piles																																														
10	Pour piles																																														
11	Pour wall foundations																																														
12	Cure wall foundations																																														
13	Concrete old cesspool																																														
14	Grub out old foundations																																														
15	Lay hardcore bed																																														

121

6. Site Organisation

SITE SECURITY

Hoardings and fencing

Hoardings and fencing should be planned carefully in order to achieve the required service from them at the minimum possible cost.

It is necessary to obtain permission from the planning authority before erecting a hoarding and essential to erect it in the position stated in the application for approval. Similarly, the type and materials of construction mentioned in the approval should be used. Some or all of the following factors should be considered before selecting the method and materials for perimeter fencing:

(a) the security risk for the various possible forms of fencing, in a built up well-lit area, 'open' fencing may afford greater security than closed.

(b) the risk to the public if a 'close' type of hoarding is not used.

(c) the restriction on site production (if any) which a 'close' type might have.

(d) the need to present a good image on city or town centre sites.

(e) the duration of the need for a hoarding.

(f) the cost.

It should be noted that double hoardings with a space of one metre between (the space itself being filled with coiled barbed wire) was insufficient to deter thieves operating at night on one city centre site. In contrast, many housing sites are developed on rural sites without any boundary fencing. Where some of the houses in rural areas have been occupied, the security problem can change, since the occupants of the new houses might require bricks for greenhouse bases, etc.

A hoarding constructed of plywood sheets provides an excellent visual screen behind which thieves and vandals can operate within a few metres of the public. This security hazard can be reduced by having viewing points in the hoarding (what the Americans call sidewalk directors' platforms). If this is complemented by arranging the site lighting system so that some lights can be left on, the would-be thief can never be sure that he is not under observation (see Figure 6.1). It is also relatively simple to make the site gates of steel angle faced with metal fabric reinforcement to further increase the chances of detection.

Figure 6.1

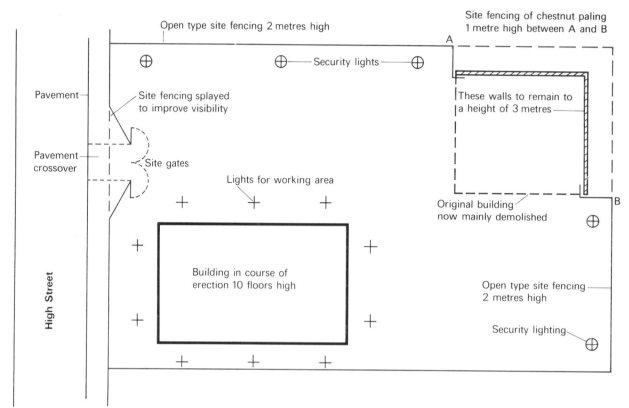

Where an existing building or wall is scheduled for demolition, the timing of the demolition may be important. A building demolished excepting for one wall retained to a height of two metres above ground level, can act as a section of boundary fence. Even when the building is not on the site boundary but within the site, the actual boundary may be marked in this section by a one metre high chestnut paling fence, whilst the retained wall provides the main security fence. It is sometimes advisable to set up an inner compound or store for particularly high risk materials.

If there is only one entrance/exit gate, the checker will be able to keep a better control on visiting lorry drivers. If his position is elevated so that he can see into the back of the delivery vehicles, he will be able to 'spot check' the incoming load and also ensure that the back of the lorry is either empty or contains only its correct load as it leaves the site.

Watchmen

Having a permanent resident watchman on duty out of working hours is very expensive and in many cases of doubtful effectiveness. It is easy for one man to fall asleep or be in a drowsy condition in the small hours of the morning. If a serious crime were contemplated, such as stealing a bulldozer, one man could easily be overcome. If there are a number of site supervisors resident in caravans on site, then a watchman with an alarm system connected to the caravans, providing they are occupied, is an improvement. It should be noted that the worst times for thefts from building sites are Christmas holidays and winter weekends when such caravans may be unoccupied. Security patrols which were all the fashion a few years ago are an easy way out but of questionable value. There is no check on visits, or indeed on whether they have even taken place. Visits are generally of very short duration compared to the period of time that the site is unmanned. Guard dogs can be very effective but their use must conform to the Guard Dogs Act 1975. The handler must be present on the site whilst the dog is being used and it is imperative that there is no possibility whatsoever of the dog escaping from the premises.

Alarm systems

Burglar alarm systems are sometimes used; these fall into two categories, those which are fixed to the doors or windows of buildings, or vibrating alarms for large unenclosed areas such as compounds. Both kinds are expensive and not unknown to break down. An alarm fixed to a door will not deter someone from breaking in through a vulnerable window. Persons engaged in burglary usually reconnoitre the site before the final act takes place.

Locks

The responsibility for locking and unlocking should be vested in one competent person. Ideally, he will live near the site and the local police will have been notified of his existence.

The key cupboard should be lockable and the number of persons authorised to hold a key should be strictly limited. In particular, keys to tool stores should be treated with special care since bolt croppers and the like may be used for 'breaking and entry'.

Figure 6.2

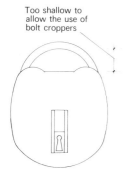

Too shallow to allow the use of bolt croppers

Padlocks should be selected with care (see Figure 6.2). It is futile to shutter windows, erect hoardings two metres high, then fit the door with a 75p lock. Similarly, there is little point in making a careful selection of the lock if the hinges can be removed easily. Hinge pins should not be extractable and screws should either be 'covered' or only extractable when the gate is in the open position.

Wherever practicable from a site layout viewpoint, site offices and compounds should be located to benefit from existing local authority street lighting, otherwise consideration must be given to leaving the office lights on at night and where the risk is high, security lighting externally. If the district is a new one to the site management, a visit to the police to discuss local hazards with the crime prevention officer should be beneficial.

The canteen is often a common target for vandals and thieves; any precautionary measures undertaken must take into account the special need of ventilating in canteen buildings, thus, if windows are to be meshed, it may be necessary to provide extraction fans.

Personal belongings and cash

After the men have reached their place of work, the drying room should be locked and if anyone requires access before the end of the working day, he should be accompanied. Alternatively, each man should be provided with a separate locker. Warning notices should be displayed absolving the firm from any responsibility for valuables left unattended on the site. Site office staff should be instructed not to leave any personal money in their desk drawers or leave wallets in jackets lying about.

Access to petty cash should be restricted to one trustworthy person and the amount kept to the minimum practical level.

Wages

The method of dealing with wages requires special care. Ideally, everyone would be paid by cheque but most building operatives would object to this and expect to be paid in cash. If the site is a long distance from the head office and wages are to be made up on site, then the responsibility for the whole procedure should be vested in two persons. They should vary the time and route of collection frequently and also change the vehicle they use. Once arriving at the site office, they should proceed quickly to the room being used for wages. Ideally, such a room would have no window and a lockable door. From that point in time, the money should not be left

125

unattended unless it is locked in a safe. When paying out on large sites, it is essential for two people to pay out so that there is always a witness and to obtain a signature for each packet handed over. It is preferable for the money to be paid through a hatch opening. Wage packets of various kinds are available which permit the checking of notes without unsealing the packet.

Security within the office

Stationery and stamps

Although ordinary writing stationery, envelopes, postage stamps, may not seem very significant from a security aspect, unless control is exercised, losses can reach alarming proportions. One person, preferably a clerk, should be responsible for the recording of all mail outward and incoming in a mail book and the same person custodian of the postage stamps.

Surveying instruments

These should not be left on site unless proper provision has been made for their custody. On large projects of long duration, it is worthwhile considering the erection, out of sequence if necessary, of any safe, small building which can be adapted for a 'valuables' storage building. If this is impractical, it is generally advisable to remove them from the site each evening, making sure that adequate insurance has been provided for their nightly storage in a private dwelling or whatever arrangement is made.

Office windows

These can be fitted with wire guards internally so that they can be opened for ventilation during the day and if lights are left on during the night, an offender may be visible to view from outside. Obviously if the building is not near the public highway, this has little value.

Materials stores

The arrangement of a materials store is described fully elsewhere. However, the main principle is that a storeman must be responsible for all goods within the stores, their receipt, checking in, for type, quality, quantity, etc, and proper, tidy storing.

The storeman would then issue stores against an official site requisition order signed by an appropriate person. This person might be a craft foreman or the general foreman or site manager. Check audits should be carried out to ensure the storeman is doing his job properly.

Material stacks should be maintained in a tidy condition and used, wherever practicable, one stack at a time so that if anyone tampers with the unused stocks, it is clearly visible. Stock levels should be minimal, and compatible with maintaining production. Stocks not being drawn upon should be sheeted up.

General precautions

Car parking

Cars should be parked in defined places and where the layout of the site permits, the parks should not be located near material stacks, stores, or the buildings under erection.

Cash sales

In some firms, the sale of small quantities of materials are arranged on Saturday mornings from the firm's yard. This can be useful in removing the temptation to steal.

Buildings

As buildings are roofed in and ready for the installation of services and finishes, they can be fitted with temporary doors. The building should be glazed throughout as soon as possible and the building may be used as a temporary store with adequate control on those who enter.

Ladders and scaffolding

Ladders not in use should be chained and locked in the compound. Ladders attached to scaffolding should be made difficult to use when the day's work is ended.

The client's premises and property

Before employing personnel in premises which are occupied by the client, known, trustworthy employees should be selected in preference to unknown newcomers. Unknown newcomers should be security checked before engagement.

Staff holidays and sickness

A proper holiday rota system should be in operation so that security responsibilities are properly transferred during staff holidays and arrangements made to cover duties during absence for other reasons.

Neighbours

If there are any housebound or retired people in the area, one can enlist their services by paying a small weekly retainer for which they will undertake to notify the police if they observe any suspicious act happening on site. If there are no residences in the immediate locality of the site, a notice may be displayed as follows:

IF YOU SEE MATERIALS, PLANT OR EQUIPMENT BEING MOVED FROM THIS SITE BETWEEN 17.30 HRS ON FRIDAY AND 08.00 HRS ON MONDAY, A ROBBERY IS TAKING PLACE. PLEASE NOTIFY THE POLICE AND YOU WILL BE REWARDED.

If anyone calls at the site and asks to be allowed to remove any plant or equipment, no matter how large or small and no matter how plausible the story told, if there is any doubt at all, a proper check should be instituted before handing over anything. Wherever practicable, plant should be placed in a locked compound at night, other plant should be immobilised when work terminates for the day, and also be properly checked before work recommences the following morning. On isolated open sites such as motorways, plant which is at work some distance from the base point can be made more secure against theft by having a number of pieces of plant chained together. Naturally, some plant would be positioned side on to end to prevent the total loss of the complete fleet in one operation!

Internal sources of theft

One of the problems facing a would-be thief is how to get the materials or component off the site without being observed. A common solution is to come to an arrangement with the driver of a spoil truck who agrees to bury the object under the muck and then arranges to deliver the stolen goods on his return journey. This is commonly known as a 'muck away fiddle'.

Vandalism

From external sources, vandalism is often caused by children or juveniles. Sometimes a carefully worded letter to the local school(s) and local residents asking for their co-operation and suggesting that offenders will be prosecuted, is sufficient to bring the matter under control. A more agreeable approach is to send a carefully worded letter asking for the co-operation of parents before any loss occurs.

The damage may be coincident with the nightly closure of a local public house or the Saturday night closure of the local dance hall. Where there is repetition in the timing of the damage, the police should be consulted or arrangements made for temporary observation.

If someone is caught 'in the act', particularly early on in the life of the contract, to pursue the matter into the courts and arrange as much publicity as possible is usually a good deterrent.

Internal sources – vandalism

When a site workman or workmen decides to damage the work, detection can be extremely difficult. If it is the work of one person, he may be mentally sick or he may bear a grudge concerning a previous reprimand. There may have been a decision about bonus, holidays, apportionment of certain attractive work, a promotion of someone, etc, which he violently disagrees with and he is emotionally immature and this is his way of showing his resentment. Looking for the cause rather than the effect may be the wisest approach. One of the great dangers here is the possibility that a number of good workmen may become upset and lose motivation if the matter is not handled with 'kid gloves'.

Company policy

If the firm's security officer's job consists of nothing more than visiting sites and commenting on the state of the hoarding and noting whether fuel stores are locked, etc, it is questionable whether the best use is being made of his time. The construction of most buildings can be divided into four phases, namely:

1. Work up to ground level.

2. The shell to roofed-in stage.

3. Services and equipment.

4. Finishings.

It is assumed that external works will be proceeding concurrently with some or all of these stages.

Before each of these stages is commenced, the security officer should visit the site and discuss the proposed security arrangements for the next stage. Most firms will want to know that some consideration has been given to the cost effectiveness of the proposals relating to security. If losses are estimated to be £50 per week and the cost of a nightwatchman with guard dog is estimated to be £200 per week, the expenditure must be questioned.

The following formal job description gives some indication of the work of a full-time security officer in the larger company. He should be able to pass on to site management the combined experience of all sites, where security has failed and the reasons for this, and also where innovations have been tested and found successful. He should prepare and circulate security check lists, supply information about the cost effectiveness of security systems, make arrangements for the supply of proper locks from a locksmith (in preference to buying 'off the peg' by individual sites), and be involved in establishing a proper system of marking all plant, etc, to provide a means of identification in the event of it being stolen. Perhaps the most important aspect of his work is the help he can give to training site staff in security consciousness and to develop a total team effort to minimise waste from this cause.

Security officer   *General job description:*

1. Main efforts of the security officer are directed towards the prevention of theft. Four main groups for consideration are:
   (a) persons not in the employ of the company and not engaged on site.
   (b) persons not in the employ of the company but having legitimate reasons for being on site.
   (c) persons in categories (a) and/or (b) acting in collusion with company employees.
   (d) company employees.

2. Minimise losses due to carelessness.

3. Guard against damage by vandals.

4. Give special consideration to the site according to its particular vulnerability, possibly because of site locality, or the site lay-out, type of construction, or mode of employment of labour.

5. After the tender stage survey of the site and study of the site lay-out, establish procedures and routines for:
   (a) receipt and checking of incoming vehicles.
   (b) checking of outgoing vehicles.
   (c) limiting authority to sign for goods incoming.
   (d) limit authority to sign internal 'Requisition Notes' authorising goods to be drawn from stores.
   (e) variable procedures for the receipt on site of wages and the payment of same.
   (f) checking procedures by foremen.

6. Make special provision for difficult periods such as Christmas holidays.

7. Make adequate arrangement for the security of workmen's tools as outlined in National Working Rule 3F.

8. Where circumstances seem to justify it, make arrangements for weights and measures checks as appropriate.

9. Arrange proper lock-up for fuels, in particular petroleum spirit and set up a tight control system for its issue and justifiable usage.

10. Educate operators in the protection of their plant from vandalism.

11. Be constantly on guard for fraudulent practices at the 'clocking-in' point and stolen goods hidden under the spoil on 'muck-away' trucks.

12. Study the economics of security, ie measure losses against the cost of stopping them.

Qualifications and training

Training in the police would be most suitable.

Experience

On various sites

Special skills

1. Capable of training site supervisors, store keepers, material checkers on how to guard against theft and fraud.

2. Vigilant.

SAFETY ON SITE

Policy

In some aspects of site policy (on matters other than safety) there may sometimes be vagueness, but in the case of safety, the law is clear and precise. Section 2(3) of the Health and Safety at Work, etc Act 1974 states that

Except in such cases as may be prescribed, it shall be the duty of every employer to prepare and as often as may be appropriate revise a written statement of his general policy with respect to the health and safety at work of his employees and the organisation and arrangements for the time being in force for carrying out that policy, and to bring that statement and any revision of it to the notice of his employees.

There is, therefore, an obligation to have a formal written statement of policy and to up-date it as necessary.

How can this formal written statement become an instrument for the application of safe work practices on site? The Act also places a duty upon 'employed persons'. Section 7 of the Act reads as follows:

It shall be the duty of every employee while at work:

(a) to take reasonable care for the health and safety of himself and of other persons who may be affected by his acts or omissions at work; and

(b) as regards any duty or requirement imposed on his employer or any other person by or under any of the relevant statutory provisions, to co-operate with him so far as is necessary to enable that duty or requirement to be performed or complied with.

Section 8 places a duty on all persons, whether they be employers, employees or self-employed, and states:

No person shall intentionally or recklessly interfere with or misuse anything provided in the interests of health, safety or welfare in pursuance of any of the relevant statutory provisions.

It is abundantly clear that the Act spells out a message of total involvement and commitment of everyone at the place of work. In achieving the desired ends, the construction industry as a whole faces many difficulties, some of these being peculiar to the industry. For example:

1. The majority of labourers have not had formal training in the tasks which they carry out. It is easy to assume that because a man appears competent that he appreciates all the dangers associated with what he is doing.

2. There is still a good deal of casual employment in the industry. Workmen from other industries are attracted in during the boom periods and in some cases leave the industry after a relatively short stay. There are many workers just passing through.

3. The industry employs immigrant labour with attendant communication problems.

4. The process of building is being increasingly mechanised and accelerated. For example, concrete can now be pumped at a speed which outstrips in almost all cases the conveyance of concrete by crane, but how

much is really known about the arching effect of concrete being placed by the slower conventional method? How can the required difference in formwork stability be assessed?

5. Sub-contracted gangs of men work side by side, each carrying out specialist processes. Often the workman understands the dangers associated with his own work but is ignorant of dangers associated with some of the other tasks being carried out in the vicinity of his work place.

6. Incentive bonus schemes, which are generally favoured by employers as a means of improving productivity and by operatives as a means of increasing their earnings, encourage workers to look for short cuts.

In attempting to offer solutions to these points, the first three highlight the need for labour stability and adequate training facilities within the industry and individual organisations. Point number 4 calls for detailed pre-planning of complex or fast processes. Item number 5 points the finger at site managers regarding the adequacy of their communications and control procedures. In point number 6, instead of paying a man, say, £500 for fixing some asbestos sheeting as vertical cladding, the 'build-up' of the task statement would show:

For fixing the asbestos	380.00
For making the access and scaffolding according to the method statement	70.00
For wearing the mask and carrying out other described precautions	50.00
	£500.00

By this method an operative would be paid the full amount (£500) only if he works safely.

This latter point raises the question of the need for the estimator to price the work in the estimate to company approved methods. In the event of the work being new to the company, or the site environment difficult, then the company safety officer should be involved at the pre-tender stage.

Safety as practised by the site manager

It is essential that the site manager is in complete control of his site from day one and that he approves of all the methods of work which are being carried out by directly employed men and that he does not disapprove of any methods being used by sub-contractors. A high standard of general housekeeping is essential and clearly within his control, checking amongst other things that each trade is clearing up as the work proceeds and not allowing walkways, etc, to become dangerous. This usually involves the need for good communications to ensure that all trades know what is expected of them.

Three-part safety check-list

It is wise to have a three-part check-list for all major activities such as demolition, excavation, formwork, concreting, scaffolding and brickwork. The three parts concern,

(a) before the work commences,

(b) during the conduct of the work,

(c) after its completion.

The following is an example for excavation work

Before excavation

1. What is the proposed method and sequence for the work?
2. Is the proposed timber/sheet piling suitable for the job, and available in sufficient quantity?
3. Have the locations of all existing services been ascertained and communicated to those concerned?
4. Is all the equipment now available including
   (a) temporary traffic lights and generator
   (b) fuel for generators
   (c) stop/go signals
   (d) flashing warning lights
   (e) batteries
   (f) barriers and disruption notices
   (g) trench ladders
   (h) bridges for access across trenches
   (j) wire bonds for temporary support of existing cables where exposed in trenches
5. Is all the required plant available?
   (a) is there a pump?
   (b) spares and ancillaries for plant?
   (c) fuel for plant?
   (d) pump to transfer fuel from barrel to plant?
6. Are the men selected for the key roles all competent?
   (a) ganger
   (b) timberman
   (c) plant operators
   (d) banksmen

During excavation

1. Is the traffic being properly controlled?
2. Is there any danger to pedestrians?
3. Is 'muck' being carried on to the road or footpath?
4. Is the trench timbering being done properly?
5. Are the spoil heaps being properly controlled?
6. Is there anything that could fall into the trenches?
7. Are trench ladders well positioned and being used?
8. Are there any trenches without ladders?
9. Have bridges got guard rails?
10. Do any trenches have to be jumped?
11. Have all exposed services been properly supported?
12. Have cable cover tiles been carefully and safely stacked away for re-use later?
13. Is pumping activity drawing material from behind the timbering?
14. Are machines working too close to each other?
15. Have spoil trucks got adequate access to the machines?
16. Are site dumpers being used safely? Are efficient stops positioned to prevent dumpers overturning at the edge of the tip?
17. Is the back filling commencing as soon as possible?
18. Is back filling well consolidated?
19. Is tidying up and sweeping following on immediately after back filling?

On completion

1. Is defective timber withdrawn from use?
2. Plant and equipment properly checked over and brought up to standard?
3. All spoil removed?
4. Pavement made good?

5. All notices and barriers removed?

6. Any excess materials (pipes, etc,) returned to store point?

7. All new installation properly tested and approved?

The involvement of all personnel in site safety

Although the site manager is responsible for the overall control of safety, he will undoubtedly want the topic discussed at the site meetings and would expect a useful contribution to current safety matters at the weekly meeting of foremen. The degree of response which the operatives give will depend in the final analysis on the example set by the supervisors. Example is the most potent method of forming habits and traditions. In construction, this can be achieved by:

1. Being seen to plan the layout of work places with safety in mind.

2. Developing, using, and maintaining safe methods of carrying out work.

3. Providing the right type of plant and equipment.

4. Exercising care in the selection of personnel for the responsible posts, ie, banksman, timberman, etc.

5. Arranging for the presence of trained operators for posts such as crane driver, fork lift truck driver, hoist operator, dumper driver, etc.

6. Carrying out site inspections and examinations in an efficient manner and not just complying with the clerical procedure of signing the register.

7. Arranging for the availability of protective clothing, etc, and for its proper storage, distribution and collecting in.

8. Installing proper systems for the maintenance of plant and equipment, arrangements for spares, storage of fuels, etc.

9. Arranging for the provision of fire fighting equipment appropriate to the work being done.

10. Communicating well, not only through the compulsory display of statutory notices but also warning workmen of potential hazards on site.

The site environment

A clean, well-ordered environment will tend to give rise to constructive, helpful, and co-operative attitudes towards safety. A necessarily dirty job is no excuse for untidiness or unsafe practices. (It is very important to dispel the attitude of the rough, tough builder who thrives off danger.) This type of environment will have:

1. Clean canteens and toilets.

2. Clean walkways about the site and in the partially completed building.

3. Good artificial lighting when artificial lighting is necessary.

4. Temporary electricity supplies and compressed airlines will not be draped about dark stairs and passageways, or in the mud, but will be secured in a tidy and safe manner.

5. Managers will train operatives in operations which are new to them.

Enthusiasm of all personnel

By harnessing group involvement, team loyalty is fostered and group safety is assured.

Despite the positive approach, managers should not be afraid to discipline operatives for breaches of safe work practices.

In accordance with the Regulations, safety supervisor(s) should be appointed, the wise site manager will work with them and through them, never doing anything which undermines their/his position. Safety supervisors should be given every opportunity to keep up-to-date with modern techniques and aids to good practice. An up-to-date file of all appropriate regulations should be maintained on site.

SITE LAYOUTS

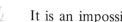

It is an impossibility to adopt any one set of standards for the manner in which to lay out a site. There is clearly a close dependency between the three factors, method statement – site layout – overall programme. Some factors for consideration:–

1. Relative size (area of space occupied) of proposed building and the area of space left to work from.

2. The volume of building compared to site area.

3. The quantity of work below ground level.

4. The site location (City of London or rural Somerset).

5. The site environment (meadowland or large hospital neighbourhood).

6. Whether the site is level or sloping, the nature of the soil, obstructions in the ground such as old foundations, etc.

7. Access and egress conditions.

8. The type of construction which includes consideration of the degree of standardisation in material selection.

9. The nature of the materials. For example, ready-mix concrete compared with pre-cast units or site manufactured components.

10. Material availability and the confidence with which this availability can be predicted.

11. The contract period and rate of production envisaged.

12. The time of various parts of construction relative to the seasons.

No doubt there are other equally important considerations.

Generally, material supplies stored inside the shell of a building in course of erection may:

1. Create the need for extra propping to suspended floors.

2. Interrupt the programme of work and delay contract completion.

3. Expose the materials to damage or theft by the site workmen.

4. Complicate the procedure of material issues and stock control.

Many of the above points do not apply in the case of shop premises when the main contractor's work often does not include the finishings, these being carried out by the lessee according to individual requirements.

Site layout is very much influenced by type of construction, density of development, and whether the site is level or sloping. Low rise, low density housing on a level site with many dwellings probably warrants the use of a fork lift truck. Likewise it is often impossible to erect a tall building on a compact city site without some type of tower crane.

It is desirable that some materials must be lifted immediately from the delivery vehicle into or near the final fitted position. This could include:

(a) ready-mix concrete
(b) large heavy pre-cast concrete units
(c) the lift cage or escalator units

It should be noted that a hoist cannot unload a lorry, whilst a crane can.

Retentions and cranage

Materials or components delivered to site but unfixed will normally carry some kind of retention which may influence the timing of the delivery. If a material or component must be unloaded by crane and the plan is to have cranage on site for a specified period, obviously maximum deliveries must be accomplished whilst the site crane(s) are available.

If on the other hand it is planned to 'hire-in' cranage on a few specified dates, access must be maintained for the crane vehicles and perhaps more important, space for the jibs to operate.

The decision on the last point may be influenced by load weights and fixed crane locations. For example, if there are 500 units in the building each weighing 2 tonnes and 3 units each weighing 20 tonnes, normally the site cranage will be selected to deal with the 500 units and 'hire-in' for the 3 units.

Figure 6.3   **Locating site offices**

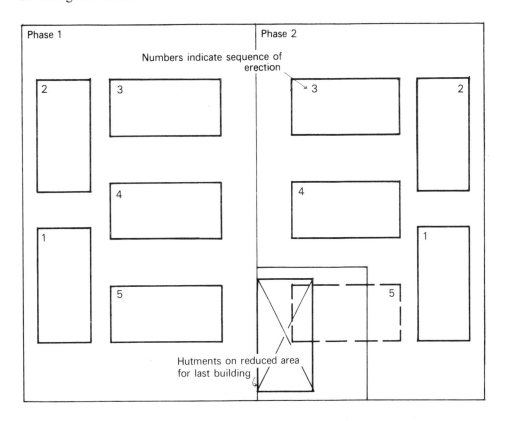

135

Planning the layout

Moving buildings or stores areas during the tenure of the contract is often costly and should obviously be minimised. Yet the site layout requirements will change considerably as the work progresses.

Siting hutments and stores

Before any hutment or stores position is decided upon, it is worth considering the following questions:

1. What ultimately is the use of this space?

2. When must it be vacated to satisfy the needs of the building programme?

Before commencing a building project, it should be visualised in four or five stages of completion and an attempt made to assess what would be an ideal site layout for each stage.

From this study a compromise location can be established for all hutments and installations, such as temporary service and distribution lines, which are planned to have a 'once and for all' location on site. Many such positions will be a compromise between what is desirable and what is possible and the question of cost effectiveness will arise.

It is desirable that the canteen and toilets must be as close as possible to the area in which the majority of men will be employed. For example, if the gross cost of employing a man is £24 per day and a man on ordinary rates might walk at a rate of 3 kilometres per hour, then assuming an 8 hour day the man is paid £1.00 to walk one kilometre. Assuming that the canteen is 100 metres from the work place, the following might apply for each operative:

Walking from drying room to work at 08.00 hrs	100 m
Morning teabreak return journey	200 m
Mid-day break	200 m
Afternoon tea	200 m
Return to drying room at 17.00 hrs	100 m
Visits to toilet during day	100 m
	900 m

900 m x 5 days = 4.5 kilometres per week
For 50 operatives = 225.0 kilometres per week
Weekly cost of location = £225.00

Figure 6.4 illustrates a principle which may be used when similar considerations arise for goods which are being transported from the stores to the workplace area by hand.

Public nuisance aspects

In carrying out construction, care must be exercised to avoid:

1. Excessive noise.

2. Excessive dust in the atmosphere.

3. Dirt on the public roads.

4. Obstruction on the public roads.

The Noise Pollution Act 1974 is intended to protect the public against excessive noise, and permitted noise levels are stated in the Act.

Inspectors are appointed by local authorities to ensure compliance to the Act and the general case is that inspectors do not themselves seek those who infringe the Act but they await complaints by members of the public and then investigate. If a contractor is making excessive noise in an area of commercial buildings, offices, banks, and the like, the inspector will soon be informed and active. In an industrial area, a similar noise level might not provoke the same reaction. A bakery or a dairy will soon inform the authorities if construction operations are making an offensive dust.

These two aspects should be considered when deciding the location of noise or dust producing activities. If the activity cannot be moved, ie the demolition of a building, then the question of method, protection, and timing may arise.

Some methods of working are noisier or dustier than others. It can be extremely difficult to 'deaden' a noise level, therefore care should be taken when siting noisy plant. Dust in the atmosphere may be controlled by damping down or by using screens which can also reduce the noise level. A damp hessian screen will considerably reduce airborne dust. Demolition carried out in showery weather is less likely to cause offence to the public. Both from the noise and dust aspect, the concrete mixer should be set up on the opposite side of the site to special neighbours such as hospitals.

Figure 6.4   **Optimum position for least cost of walking to one central point.**

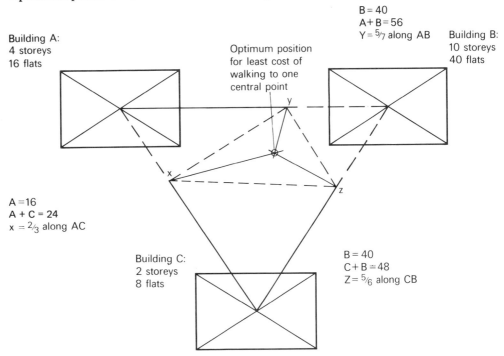

Building A:
4 storeys
16 flats

B = 40
A + B = 56
Y = 5/7 along AB

Building B:
10 storeys
40 flats

Optimum position
for least cost of
walking to one
central point

A = 16
A + C = 24
x = 2/3 along AC

Building C:
2 storeys
8 flats

B = 40
C + B = 48
Z = 5/6 along CB

Dirt on the public roads is a frequent complaint against contractors. It is usually a question of deciding whether to install a wash down point or on very few occasions directing a labourer to clean it up immediately.

The most common form of obstruction on public roads by contractors is the rubbish skip. The legal aspect regarding these is stated in the Highway Act of 1971, Cl 31.

Buying space   Quite frequently on town centre sites, the site space is very often inadequate to conduct operations from. If it is impractical to operate from within the partially completed building, it will be necessary to either conduct some of the operations at the contractor's headquarters or acquire a temporary site for the purpose. This will give rise to the problem of transporting formwork or joinery, steel reinforcement or fabricated steel, etc, from the place of assembly to the site. The use of standardised containers, pallets, and skips, may be necessary, and requirements of off-site sub-contractors would obviously need consideration. On such sites the storage of formwork temporarily out of use (propping, scaffolding, etc,) can present major problems. Lower levels of scaffolding below the current construction level can sometimes be utilised, but obviously the question of the safety of personnel working at ground level is of paramount importance.

Where temporary toilets can be connected to the local authority sewer and supplied with temporary water source for flushing purposes, there are obvious advantages. Chemical closets can be readily moved about to be near the workforce and minimise walking time. Trailer mounted toilets and wash places are clearly very useful on widespread sites to move about as the area of concentration of work moves.

Problems with tall buildings

When tall buildings are in course of erection, there comes a time when it is economic to arrange toilet facilities within the building. If the height of the building, and the number of men employed, is insufficient to justify the provision of a passenger lift, efforts must be made to minimise the amount of time men spend clambering up and down ladders or staircases to use the welfare services. However, when toilets are placed within the building, extra special attention should be given to hygiene. If mid-morning and afternoon tea is taken within the building, the immediate area should be made as congenial as circumstances permit.

ADMINISTRATION

Site administration is a very important part of site activity. Unless adequate site records are maintained, serious problems arise. There must be continuity in records, particularly with regard to factors which affect payments. Since measurements affect payments, some written records may require supporting with photographs if they relate to work which is likely to be buried or in some way inaccessible before agreement on measurement has been reached between the client's and the contractor's quantity surveyors. Records on any site must be maintained according to the firm's system. A site manager expects guidance in setting up the site administrative system (many firms produce a manual) and in smaller organisations, the contracts manager would assist the site manager to do this. Not all site managers or surveyors see the contract through from start to finish, so properly minuted site meetings, signed daywork records, and a methodical approach to variations accounts are prerequisites of a satisfactory financial conclusion to a contract. Apart from the aspect of profit, adequate records are important from many other viewpoints. Amongst these are:

1. Evidence that disciplinary action has been properly carried out.

2. In the event of a dispute with the client after the completion of the contract.

3. Queries being raised by the Inland Revenue after completion.

4. Evidence to prove which operatives worked on the various processes may protect the contractor's position in the event of compensation claims.

Site administrative forms

The first need with information handling is the design of the basic system, ie who needs to know, and this leads to form design, what they need to know. Perhaps the next consideration is how and when, and how frequently the information will be transmitted.

Thought should also be given to deciding how this information will be filed on site. Some estimate needs to be made of how much file capacity for each item is required. One would hope that one accident book and one register for inspections would last a good while but 'goods received' sheets soon assume considerable thickness.

Many firms use box files each containing four or five manilla pocket folders, each folder having a capacity of about 20 mm thickness. Different colours can be used for the various divisions of the system, ie plant, materials, labour, production. Carefully set up systems minimise on the use of time, one of the most costly resources in all work.

Concerning labour	(a) new employee
	(b) time sheets
	(c) notice of termination of employment
	(d) warning of misconduct
	(e) dismissal
	(f) labour allocation sheets
	(g) daywork sheets
	(h) labour returns (fluctuating contract)
	(j) accident report
	(k) days off required
Concerning plant	(a) request for plant
	(b) request for transport
	(c) plant time sheet main contractor owned
	(d) plant time sheet plant hire
	(e) plant transfer notes
	(f) plant breakdown and faults
Concerning materials	(a) materials requisition
	(b) materials progress
	(c) materials on site at valuation
	(d) materials transfer
	(e) materials returned to supplier
Concerning contractual aspects	(a) confirmation of site instruction
	(b) record of extra work by sub-contractor
	(c) daywork sheets
	(d) record of extra work by sub-contractor
Miscellaneous	(a) theft
	(b) insurance and police
	(c) inland revenue 715 tax receipts
	(d) negative performance of sub-contractor
	(e) request for overtime permit
	(f) petty cash
Production	(a) weekly production report
	(b) site diary
	(c) production delay imminent

The telephone

Although all of these are shown as 'standard forms', in fact many would be used simply to place on record what had already been transmitted verbally on the telephone. There must be efficiency in using the telephone. The first aspect must be restricting the use of the telephone to the firm's business excepting for real domestic emergencies. Whenever it can be justified, a coin box telephone should be provided for the use of sub-contractors and labour. This box should be outside the offices but in view from them.

When telephone messages are received, they should be noted on a proper message pad. A proper system should be instituted to ensure that they are received by the person intended; scraps of nondescript paper are easily lost.

SITE COMMUNICATIONS

In order that the project proceeds smoothly and is brought to a satisfactory conclusion, the channels of communication between the parties should be established and understood by all from the outset. Someone must accept the responsibility of ensuring that the channels are established and also being used. In many contractual arrangements, this responsibility falls to the architect, although in some cases, (eg package deals) it can fall to the main contractor. Once the project is under way, there can be three focal points for communications:

Project communications	:	The Architect
Interdepartment	:	The Contracts Manager
Site	:	The Site Manager

It must be stressed that these are all part of the same body and not three separate systems. It is the intention here to concentrate on site communications, not the other two points. The following general areas are worthy of consideration:

1. Scheduling and calling forward of materials.

2. Communicating with sub-contractors' head office.

3. Communicating with sub-contractors' supervisors on site.

4. Giving instructions to directly employed men.

5. Arranging and attending co-ordination meetings.

6. Planning and preparing production programmes.

7. Reading and understanding drawings and specifications.

8. Preparing drawings and sketches.

9. Writing letters, reading letters.

10. Self communicating – thinking.

11. Completing and sending to Head Office various production and financial returns.

The foregoing eleven items could easily account for 90-95% of a site manager's total time, thus showing that his role is mainly as a communicator and how important it is to develop skills in this.

The first skill must be recognising which component parts of each of the eleven items can be standardised and therefore establish a routine. The next point to consider is whether the whole items or a part can be delegated, and if it is deemed desirable to delegate, to decide to whom it should be delegated.

Scheduling and calling forward material

A general foreman can be expected to call forward materials; preliminary schedules are often prepared by the buyer. In calling forward materials, it is necessary to give consideration to the supplying organisation's problems. A reasonable period of notice should be given; this time allowance can vary considerably according to changing demand conditions. A monthly bulletin prepared by the buying department and circulated to sites, is useful in keeping site management 'tuned in' to changing material market conditions. The surveyor will prepare a monthly valuation on materials comparing amount measured against deliveries and stock. This should point the finger at waste, whether caused by lax security, or carelessness in use. A part of the task of calling forward is to check that the amount in question is what was expected at this stage of construction and avoid the disappointment of finding that materials required are 20% in excess of the original materials plan.

Communicating with sub-contractors' head offices

When the buyer negotiates with sub-contractors, he will obtain from them information usually by a standard form, stating how long they estimate the work would take to complete, the labour strength they could provide for the work, the earliest date they could start on site, and the period of notice they would require before moving on site. The site manager will operate within these agreed conditions unless they conflict with the contractor's programme, and usually arranges a preliminary meeting with the sub-contractor's contract manager to discuss the approach to the work, storage arrangements, accommodation, etc. The sub-contractor's contract manager should be invited to the monthly site meeting which precedes his arrival as a part of the site team and at all subsequent meetings until his contribution is complete. Any telephone conversations taking place should be confirmed in writing by the site manager and a copy retained; the great value of the site diary giving a reference to the filing of these copies cannot be overestimated.

Communicating with sub-contractors' supervisors on site

The sub-contractor's supervisor should attend the weekly production meeting, when all supervisors should be encouraged to contribute towards the tactics for the coming week's work. Apart from this, the general foreman will be in frequent contact with the sub-contractor's foreman. The most important aspect is to preserve a good working relationship so that common objectives can be achieved. Often, the most delicate points of verbal discussion are work quality standards, production delays due to poor attendance or undermanning by the sub-contractor. These aspects demand considerable tact, may demand considerable firmness, and always need care in the use of language. The latter point is a weak area with some general foremen.

Giving instructions to directly employed men

The great problem can be the wide variation in background of the labour force. More and more one finds university undergraduates (during holiday periods), Commonwealth Immigrants, Irishmen and UK labourers operating on any large building contract. Traditionally, in the construction industry instructions must, at grass roots level, be given verbally to be acceptable. It is wise to recognise three stages in giving a verbal instruction:

(a) stating the instruction

(b) finding out if it has been understood.

(c) checking that it is being complied with in the manner anticipated.

All this must be achieved without appearing to be pedantic. Finding out if the instruction has been understood requires the supervisor to give the operative(s) the opportunity of speaking in order to assess his/their appreciation of what the object of the exercise is. The major role of the supervisor is to guide and direct the efforts of workmen towards common goals. In achieving these goals, he will often have to adopt the role of teacher without appearing to teach. To say, in effect, "Let us work through the manufacturer's instructions together" appears much less egoistic than saying, "I will show you how to do it".

In checking that the instruction is being complied with, it is advisable to do this when the operatives have had time to start but are not too far advanced, to minimise the cost of, and demoralisation caused by, having to alter work.

Arranging and attending meetings

The types of meetings held on site and their frequency varies between organisations and the particular construction needs. Two types are considered here:

1. Meetings called monthly by the contracts manager and attended by:

   (a) the clerk of works
   (b) site manager
   (c) general foreman
   (d) contracts managers of the sub-contracting firms at work on the site at the time.
   (e) site buyer
   (f) secretary

2. Meetings called weekly by the site manager and attended by

   (a) the general foreman
   (b) materials co-ordinator
   (c) planner
   (d) site trades foremen
   (e) sub-contractors foremen

Meetings are set up in order that people involved in a common project can co-ordinate their activities, so that each and all can measure the performance to date against the anticipated achievement by this date.

Monthly meeting of sub-contractors

It is worthy of mention that if these meetings are to be successful, the members of the meeting should be protected from interruption by the telephone or other outside influences. Secondly, there should be a room which is large enough to permit the proper conduct of the meeting. This might include displaying drawings or programmes for all present to see. If a person is required to attend the meeting for a short period of time, ideally the item on the agenda with which he is concerned should be dealt with early so that he can leave after this and thus avoid spending time as a captive spectator. If logical sequence of items for discussion does not permit this, the chairman should be able to give people with a limited contribution to make some idea of the time scale for the meeting.

It is important that a proper record is kept of all the discussion and decisions taken and published as soon after the termination of the meeting as possible.

Conduct of a site meeting

A meeting is unlikely to be successful unless the chairman and secretary do their preparation properly. Before the agenda is prepared, the site manager should be considering what items need to be discussed, when they need to be discussed and the possible solutions to the problems posed. For a meeting to be successful, the first prerequisite is a good chairman. In the case of a sub-contractor's/craft foreman site meeting, the chairman has a particularly difficult task. It is usually not possible to put an upper limit on the size of the meeting for obvious reasons and he cannot choose the representatives because their supervisory role automatically selects them.

The purpose of such a meeting is to ensure that all work is proceeding efficiently towards a satisfactory completion. The chairman should have worked out, prior to the commencement, a projected time allowance for each agenda item.

Site meetings are in many ways a very expensive way of co-ordinating the activity of the site work force. It is therefore essential that the meeting is conducted in an orderly, positive, brisk, and fair manner.

This presents a difficulty, permitting and encouraging participation, but detecting and stopping irrelevant discussion. The chairman should, either by words or attitude, compliment those people who have clearly done their preparation for the meeting. Each agenda item discussed should be completed before the next point is commenced. If certain action is necessary as a result of the discussion, the person(s) involved should know precisely what is expected of them. If there is a clerk of works appointed to the contract, he would be in attendance at these meetings in his capacity as "an inspector on behalf of the architect". The clerk of works would naturally be interested if changes of method of working were under consideration since these may have a bearing on the quality of work.

7. Materials Supply

Some historical considerations

It is impossible for anyone to say when the de-skilling of the construction industry began but the movement was certainly well under-way in the late nineteen-fifties and early nineteen-sixties. Complementary with this movement, systems of prefabrication were being developed. Prominent amongst these were light latticed metal framed systems particularly suitable for low rise buildings and some specifically designed and developed for school building.

Another innovation was the development of high rise dwellings of all concrete construction, some of which had in-situ concrete frames, with floors, walls, staircases, balconies, etc, of large pre-cast units. Other systems were composed entirely of pre-cast concrete. Following closely behind these was a growing interest in timber framed house construction. It was a period of great interest in constructional developments. Fewer sites for development were available than in former periods, and of those that were, many were far from ideal, from both designer and builder viewpoints.

Not unnaturally, the cost of land was rising as a percentage of total building costs. These changes created new problems and new opportunities both for designers and contractors, although they were not necessarily the same problems and opportunities.

Designers were faced with designing in new materials of which they had limited or no previous experience. In some cases, the materials imposed new design restrictions or gave new freedom. Modular co-ordination, thermal and sound insulation, compatibility between old tried materials and relatively new ones, weather proofing and jointing problems, all had to be researched.

The builders' problems were equally numerous. The increase in land costs caused clients and architects to look for increased density of site development, thus leaving the contractor less operating space in which to handle larger units and to construct a usually taller building. An added dilemma, from a site manager's point of view, was the disappearing craft force. This meant a loss of unofficial group leaders who had been trained to measure and handle materials and components with some care; pride in workmanship was certainly not an ascending philosophy. There was a need for an increase in the ratio of site supervisors to site labour.

There was a considerable growth in the volume of sub-contracted work creating more problems of co-ordination. Coupled with this, the site manager required better support in the form of improved documentation from the design team, more detailed pre-planning before the occupation of the site, and a more comprehensive back-up service from his head office concerning the supply of materials, labour and plant.

By this time, the tower crane was a familiar sight and excavating and earth moving equipment were both well developed. There were, however, serious shortcomings in plant suitability for many of the new processes. The nineteen seventies have been noteworthy in the concentration which has taken place concerning materials movement, packaging, handling, etc. Many ideas and techniques have been 'borrowed' from other industries; some ideas which have been developed in Europe or America are now being practised here in the UK.

It is inevitable that some of the smaller builders are unable to alter their methods and keep the use of plant and equipment in step with larger contractors. Palleted or packaged bricks, and pumped concrete are inappropriate for most jobbing work.

Thus the knowledge and training required for operatives engaged in repairs and alteration work is different from that of the operative engaged in plant-intensive operations on new large construction work. In some ways the industry is more fragmented than ever before.

The buying function

The buyer, or in his more modern role, the materials controller, is concerned solely with the procurement, storage, handling, and use of materials.

At the tender stage of a contract, the estimator is often responsible for raising the initial enquiry with the potential suppliers of materials. There are strong arguments against this. Many people think that this should be done by the buyer. The buyer is in a much better position to know supply and price conditions, which items to obtain quotations for and which materials he can assume prices for. The buyer will have more knowledge concerning the capabilities of suppliers and sub-contractors.

The materials controller

On large works there is a case for those who have responsibility for the buying function to be involved in the pre-tender planning processes, giving their expertise on the timing of arrival of materials, access and site road problems in relation to delivery vehicles, packaging, handling arrangements, equipment and plant, site storage, security and protection of materials. If the contract is won, then such a person can move to site as a site based materials controller.

Records of suppliers and sub-contractors

A good buying or materials control department will create and maintain comprehensive records concerning suppliers and sub-contractors. This can be done by arranging a feed-back system from the site based materials controllers.

It is wise to consider what work the sub-contractor is equipped to do and to recognise that over a period of time, his managerial skill may improve or diminish. A contractor may operate over a wide area such as the South East of England or the North West of England. A sub-contractor may be competitive in part of such a region but not all of it. Competition is not necessarily equal throughout the region and the buyer should have knowledge of each sub-contractor's area of competitiveness. This is useful when preparing a short list for quotations. During the course of a contract, a similar assessment should be made of the materials suppliers' performance.

Standard of communications with potential suppliers and sub-contractors

It should be mentioned that whether the firm obtain good or bad suppliers may, to some extent, be related to the quality of communication which the supplier receives from the contractor's buyer. When a supplier is invited to quote, he should be informed of the type of contract for which the quote is requested. He will want to know whether 'fluctuations' apply or whether it is a 'fixed price' contract. If it is a 'fixed price' contract, he will want to know when the materials are required and the date for completion of the contract.

Many contractors have a set of conditions applicable to suppliers and these will be included with the 'enquiry' and specifically referred to in the enquiry letter. It is usually necessary to send a photostat copy of that section of the Bill of Quantities which applies. One copy of the Bill to suppliers and two to sub-contractors is the normal arrangement. Suppliers like to send their quotations out on their own stationery with their conditions of supply printed on; sub-contractors retain one copy of the 'enquiry' and complete the other and return it to the contractor as their quotation.

Particulars of the way in which loads are to be prepared and delivered should be stated. If materials are to be delivered loose, crated, or crated and palleted, this should be stated, and also whether cranage is to be supplied by the contractor or by the supplier. The quotation should also state if the price is based on full loads of ten tonnes, (or whatever the accepted standard is for the material) or on part loads of 3 tonnes.

The merchant should be informed if the supplier is required to produce drawings of fabricated materials, which the architect may want to examine and approve before fabrication commences. It should also be stated if samples of materials are required for the architect's approval. Should specified materials prove to be unavailable at the time required, guidance should be given to the supplier on the submission of particulars of alternative materials.

Liaison with the design team

At this stage, queries may arise when the buyer is 'taking off' from the drawings and comparing these quantities with the 'billed' items. Arrangements should be made so that the buyer may telephone to the architect or a named assistant and obtain clarification concerning such queries and thus be able to notify suppliers and potential suppliers without delay.

The architect should be asked to confirm that all nominees for the supply of PC items have been advised of their liabilities under the main conditions of contract.

Nominated suppliers and sub-contractors

When the buyer receives from the architect quotations from nominated suppliers or sub-contractors, it is essential that these are the original quotations and not photostat copies, so that the buyer is absolutely sure that he is in possession of all the conditions of sale. These should then be scrutinised and if there is doubt about any points, the queries should be raised with the architect immediately.

Table 7.1 shows a suitable format for the comparison of quotations from suppliers.

Table 7.1

Quotes comparison

Supplier	Date received	Whether quoted right	Total price £	Discount	Fixed Price	Conditions & remarks
Jones Bros.	18/8/78	Yes	8040	2½% cash 17% trade	3 months	His Clause 8, repeated here, not acceptable

Use of decision rules

In recent years, there has been increasing use of quantitative techniques in the preparation of tenders. These fall into three categories.

1. Bidding strategy. This is an attempt to determine at what sum the contractor will just be awarded the contract. (See Chapter 10).

2. An attempt to forecast material price stability.

3. Quantifying the process of supplier selection.

Perhaps the most significant feature of the post 1970 era is the high level of uncertainty; this applies in particular to commodity prices.

The number of organisations which study future trends, and pronounce on possible outcomes, is only equalled by the number of different answers which they give. Until the environment stabilises, it seems unlikely that a buyer operating at 'firm' level can succeed in predicting future price levels, when government, bank, and industrial economic forecasters may fail. All buyers should be alert to the available sources of market intelligence, should recognise that some commodities have greater price predictability than others, and use the knowledge accordingly. Two examples of this follow:

Assume a major supplier of bricks digs clay within one hundred miles of the centre of a large conurbation where ninety percent of his products are sold. He uses all his own directly employed labour and plant to 'win' the clay, make the brick, and transport the brick to the user. The past price performance of the brick in that region should be considered and a prediction made for the future.

Consider now the supply of copper products. Whilst the winds of change are blowing across Africa, he would indeed be a wise man who could talk convincingly of a 30% chance of something happening to copper prices. It is obviously wise to relate a price policy to uncertainty, but be sure when attempting to quantify uncertainty that ideas are based realistically.

Predictors

The National Federation of Builders' and Plumbers' Merchants are a very well organised trade association and it is suggested that buyers would do well to refer to them, if in any doubt when predicting future prices.

Materials Handling

It is possible to consider materials handling from at least five viewpoints, and each of these can be sub-divided into different aspects.

For example, the following division could be considered:

1. Manufacturers of building products.

2. The builders merchants.

3. Large civil engineering works.

4. Repair and alteration work.

5. Large new construction work. (To be discussed in greater detail below).

The manufacturer   A manufacturer of concrete cladding panels may set up his organisation to supply individually designed high quality concrete cladding panels. Since the products will vary considerably in weight, size, and shape, the lifting of any one panel onto a road vehicle may present unique problems. More typically, a manufacturer is concerned with maintaining an even rate of production. To achieve this, he needs a standard product design or at least as few variations from standard as possible. Since the demand for building products varies, the manufacturer must stockpile. Manufacturers of bricks, blocks, and concrete products find it necessary to 'lock up' a considerable amount of capital in stock, as do the extractors of sand and coarse aggregates.

Most of these products can be stored outside on level ground. Many other products, notably those made of plaster, plastic, porcelain, glass, and most metal products, must be stored inside warehouses.

Not unnaturally, the merchant develops ways of packaging and methods of storage which are appropriate to his needs, and he buys plant to match these requirements. Sometimes his handling methods and ways of packaging are not particularly appropriate to site methods. In the absence of a co-ordinating body for the regularising of handling methods for building materials, this is not surprising.

Planned warehousing

The manufacturer, having successfully marketed a product, will develop a packaging system and then develop a warehouse to suit the system. He will maximise on the amount of space which is used for aisles. There are particular problems when, say, a fork lift truck is carrying a palleted load down a narrow aisle and wishes to change direction through 90°. Obviously, warehouse stacking must be very carefully planned. Some modern systems of warehousing are fully automated with computer controlled stacker cranes. Instructions by punch card or magnetic tape are fed to the computer in the usual way and the loading or unloading is monitored by computer.

The main point to note is that none of this packaging, preparation, storage, and re-loading back on to the suppliers' vehicles, has any relationship with the way the contractor would like the goods to be delivered to him on site. It is inconceivable that methods of site handling could be completely standardised as there are many variable factors, but it is probable that considerable advancement in this direction could, and will be made in the future. Some contractors could, with advantage, spend more time thinking about the way in which they would prefer goods to be delivered and then order accordingly, giving adequate time to the supplier. Special packing arrangements may result in the initial cost of materials being higher but cost viability would be assessed before adopting them. For instance, an organisation building houses as a speculation usually has good control over the organisation of systems of materials delivery appropriate to the construction process.

For example, assume a number of pairs of houses have been constructed up to ground floor slab stage. The remaining materials can be ordered in, say, eight packages, package one containing:

1. All ground floor window and door frames.

2. All damp proof course materials for the horizontal dpc and dpcs around window and door frames.

3. All lintels, lintel trays, and reinforcement for reinforcing brickwork over openings.

4. All first floor joists and metal joist hangers required for the temporary support of joists.

5. All nails and screws for the floor construction.

6. All bricks and blocks required for the fabric of the ground floor, and cavity tie irons for this work.

7. Cement and sand for the mortar, plasticisers and colouring agents, polythene sheeting for protection, if necessary.

The adantages of this approach are immense.

(a) low levels of site stock

(b) predictable storage requirements

(c) easier security control

(d) less danger of production delays caused by material shortages.

The supplier   Some materials used in construction are manufactured, or extracted, and supplied by the same organisation. This arrangement is typical for bricks, blocks, pre-cast concrete, loose sand, aggregates, and ready-mixed concrete. This kind of arrangement is supportive of research and development.

One of the unfortunate aspects which can occur if development is left in the hands of individual companies, as distinct from more central industry based research, is that the site plant required to move one manufacturer's product may not be suitable to handle the product of another. The idea of standard unit loads is now being given much thought.

The further this idea is developed, the easier it will become for the site to plan their activities and minimise the variety of plant and ancilliary equipment required. The unit load idea has considerable implications for manufacturers, suppliers, and building sites. Once on site there is a further distribution problem which requires consideration. For example, the London Brick Company Selfstak system enables bricks to be unloaded on the ground by the delivery vehicle and driver. These are off-loaded in units of up to 350 bricks. These units can then be lifted by fork lift truck direct to scaffold level at points where the scaffold has been stiffened to receive such loads. A Selfstak brick barrow has been developed to cope with 350 pack of bricks in 5 barrow load units. Here is an example of the supplier thinking beyond his own efficiency problems towards the contractor's site distribution problems. It would be advantageous from the building contractor's viewpoint if all brick manufacturers were acting together in the development of brick handling equipment.

Civil engineering   Civil engineering work generally involves fewer different materials and fewer sub-contractors than building contracting. However, the quantity of materials handled is usually considerable, and one of the major materials handled is earth.

Earth may be imported to site, and in these circumstances, it generally has to be paid for or it may be exported from site; in this case it may be sold or alternatively, a tipping cost may be incurred which, coupled with the cost of transport to tip, can be considerable. Alternatively, earth may be excavated and stockpiled on site for re-use later. The conveyance of earth is open to all manner of fiddles and misconceptions and it should therefore be controlled by regularised procedures. It is recommended that not only is a separate ticket made out for each load carted away but also the time of departure and return of the vehicle, from and to site, is entered on the ticket and, of course, the registration number of the vehicle. Receipts for payments made at tipping points should be expected. When earth is transferred to another contractor or sub-contractor on the same site, costs should be agreed before the transfer takes place.

Bulking of earth

All estimates of cost and time of excavating and conveying earth should take into account the bulking effect of earth. The basic factor is that earth which is undisturbed in the ground, ie as compacted by nature, will occupy a greater volume of space when dug and in a 'loose' condition. A third volume enters the picture when 'loose' earth is used to fill a trench, the trench filling being compacted by mechanical means. The three measures are:

1. Bank measure.

2. Loose measure.

3. Compacted measure.

The matter is further complicated by the fact that these three measures can vary both in volume and weight according to the moisture content of the earth.

It should be noted that in certain circumstances, earth can be 'compacted' to a higher density than it occupied in 'bank' condition. The volume which a vehicle can carry is easily determined from the vehicle dimensions and manufacturer's load size. The weight which it can carry is related to two factors:

1. Rolling resistance.

2. Grade resistance.

Rolling resistance is defined as the force that retards the motion of rolling wheels as they move over the level ground surface. This is determined by ground conditions, tyre inflation, tyre construction, tyre wear, speed and friction in the wheel bearing. Grade resistance is the force that retards the motion of a machine as it travels on a grade against the downward pull of gravity.

It is obvious that in the estimating stages of earth transportation, consideration must be given to the prevailing conditions which include the foregoing aspects, also the condition of the contractor's vehicles and the training and experience of his drivers.

Merchants' delivery vehicles

When negotiating with suppliers, there are obvious advantages to the contractor if the supplying vehicles are equipped with cranage, slings, etc, and the driver is able to unload without site assistance. This limits the number of disruptions to the site production processes.

Material stockpile areas must be chosen carefully with regard to security, minimising the distances that materials have to be carried forward over rough ground. The merchants' delivery vehicles must be able to travel to the stockpile areas. This requires consideration in the 'timing' of deliveries relative to ground conditions, marking out routes, open trenching and in particular, marking the location of trenches which have been filled in. Where temporary site roads have been laid, the use of a material such as Terram (ICI) to stabilise the road should be considered. Flexibility in the location of stockpiles should be planned if dramatic changes are likely to take place in ground conditions.

Repairs and alteration work

The extent to which a contractor carrying out repairs and alteration work can take advantage of modern handling techniques depends very much on the circumstances of the particular work in question. The easiest and cheapest method of transporting concrete for a new suspended first floor in an occupied building may be to pump the concrete. The easiest and cheapest method of transporting a hundred roof tiles from ground level to sloping roof surface might be by means of a fork lift truck if there is a single storey glass lean-to conservatory along the full elevation of the two-storey building where the roof covering is to be repaired. Plant hire and plant leasing is now available for many types of plant and it is obvious that smaller organisations would rely heavily on these services for their needs, rather than locking up capital in under-utilised plant and equipment.

Materials handling on large new construction work

Site movement of materials

The classifications of movement are:

1. Moving materials horizontally:

    (a) at ground level
    (b) on external scaffolding
    (c) inside the building.

2. Moving materials vertically:

    (a) downwards into basement excavation
    (b) upwards on the outside of the building
    (c) upwards on the inside of the building.

Classification by type of load:

(a) earth
(b) loose aggregates
(c) steel rod
(d) caged reinforcement
(e) formwork
(f) pre-cast units
(g) partitions
(h) large sheet - glass - plasterboard - plywood
(j) boilers - pumps - radiators
(k) pipes - plastic - copper - steel
(l) composite units
(m) escalator chassis
(n) glazed windows
(o) hardwood joinery
(p) roof trusses
(q) suspended ceiling track - tiles
(r) light fittings - tubes
(s) plant and equipment for use within and outside of the building

Diversity

A brief glance at the diverse kinds of materials which are used and the restrictions imposed by the building or its environment causes one to think hard about the possibilities for developing a unit load. There is obviously a limit to the amount of standardisation which can be developed in load sizes and volumes of construction materials.

152

Continuity of control

If the right materials are to be available at the right time, in the right quantity, it is self-evident that the knowledge of the contract, the information flow related to it, and the involvement and commitment of personnel must be ongoing from pre-tender planning to final account. On large contracts it would be ludicrous to employ a skilled buyer between the award of the contract and the commencement of the construction, then seriously to weaken the communicating link between the buyer and the site when the critical materials delivery period commences.

Bearing in mind that material might account for thirty to forty percent of the builder's cost, clearly the materials resources is just as important as labour, plant or finance. What is therefore required on site is a materials controller. The role requires that his involvement commences at the pre-tender stage and terminates towards the end of the 'finishings' and 'services' period of the contract. The concept of developing the role of the buyer into that of materials controller increases the number of personnel involved with management and obviously overhead costs.

Assume a contract worth £2 000 000 to be completed in one year has contract costs and planned profit as follows:—

Contract sum	2 000 000	
Profit 4%	80 000	
HO Overheads 12%	240 000	
Labour 30%	600 000	
Materials 30%	600 000	(Wastage of materials
Plant 20%	400 000	£60 000 at 10%)
Prelims 4%	80 000	
	£2 000 000	

A fifty percent improvement in the wastage occurrence is within the bounds of possibility if the buying function and materials control function are united and given their rightful place in the hierarchy of construction management.

Gross saving	£30 000
Cost of material controller	£10 000
Net saving	£20 000

The net saving represents an improvement of 25% in the profit performance. Unless a site manager has had the benefit of a completely co-ordinated management traineeship, he is unlikely to have the right sort of background and knowledge of suppliers, supply contractual conditions, and buying generally to do the job of site materials management well, hence the emerging role of the materials controller with a number of large contracting firms.

Duties of the materials controller

At the pre-tender stage

The materials controller should advise contracts management of the supply conditions prevailing and of innovations in plant and handling equipment. Inevitably many queries are raised at this stage concerning suitability of materials, administrative aspects of payment requirements, contractual conditions, delivery arrangements, and fixing requirements. These points will be raised by the controller with either the architect or supplier, as appropriate.

Apart from the obvious involvement with materials, his knowledge of the materials in the Bill of Quantities will be useful for plant selection purposes. For example, the size of a goods hoist might be reasonably governed by the largest size of glass sheet which is specified. When it is planned to take materials from an outside storage area by site vehicle such as a fork lift truck into the centre of the plan of the building, an influencing factor on the selection of the vehicle might be the clear storey height of the ground floor of the building. If conduits, pipes, etc, are to be encased in the ground floor screed, the timing of these might be influenced by the requirements of the materials movement. The services could be positioned as early as possible so that the screed can be laid to provide a better surface for the fork lift. If this is the case, the risk of damaging the screed must be considered. Alternatively, the services at ground floor level may be delayed until after the materials movement. The stability of the ground floor slab should be considered and if it is a suspended slab, it may be necessary to route the vehicle.

For materials which have to be transferred from the fork lift truck to the hoist, the materials controller will recommend a suitable trolley so that when arriving at the appropriate floor level, the material can be conveyed directly to the point of usage. Where the load on the trolley is a single unit as distinct from bricks, blocks, or similar material, some form of lifting equipment will be required. Should this be an independent item of plant, or integral with the trolley, is a problem for the materials controller who should be a specialist in all aspects of materials.

During construction

One ongoing aspect of his work during this period is the monitoring of 'supplies in' and 'chasing' materials when production delays on site are likely to occur because of non-arrival of materials. Apart from this, he is responsible for site storage and material protection arrangements. Plant which is used solely for conveying materials from stores to usage point is often under his direct control. In order to programme the distribution of materials, it is necessary for him to maintain liaison with section and trades' foremen so that he can plan their individual material requirements. It is his duty to control administration related to materials and to measure the amount of waste which is occurring at the various points (ie delivery, storage, user, damage to fixed material).

Task objectives

Modern methods of materials control should lead to higher efficiency through better material selection, lower stock levels and thus a better financial performance. Less loss should occur by theft, wastage, damage in transportation and user waste. The ultimate quality of the work should be higher.

Advantage to suppliers

Suppliers will be dealing throughout the contract with a person who understands their way of working, normal supplier practices, and the agreed conditions of supply with regard to this contract. Suppliers should receive an improved consideration from the contractor, and the contractor an improved service from the supplier.

Palletisation of materials

The use of fork lift trucks must be synonymous with consideration of movement of materials loaded on pallets. There is no overall system for the use of pallets in the construction industry as developed or partially developed in some other industries.

Small materials

Materials such as rolls of damp proof course and cavity wall ties can easily be controlled on the pallet by means of metal cages; these cages can be removed on the scaffold prior to distributing the materials about the scaffold platform.

Loose dry materials

Pallets with sides can be utilised as an alternative to skips when the material is not required for immediate use, ie sand for use by bricklayers in sand courses at the service floor level of a tall building.

Demountable flat bed truck bodies

These systems have a tremendous impact on the suppliers' modus operandi and an equally considerable impact on site tidiness. The basic idea is that the supplier standardises his vehicle delivery fleet and purchases a number of demountable flat beds. The use of a number of flat beds to each vehicle gives the supplier a much better utilisation of his vehicles.

This system also helps the site materials controller to plan his access and storage space usage. The supplier can arrange that supplies planned for early morning delivery are loaded on to a flatback the evening before delivery. The vehicle which is to make the delivery can be engaged on other delivery work until 1730 hours on the day before delivery but can be on the road ten minutes after start time with the early morning delivery. Clearly, when a delivery vehicle has to wait to be unloaded with a vehicle driver, and primary mover idle, a 'general loss' occurs and has to be paid for, in the first instance by the contractor and ultimately by the client.

Containerised supplies

City centre developments often give the contractor little or no operating space at ground floor level. Depending upon the particular circumstances, it may be necessary to acquire the temporary use of space located as near as possible to the construction site. This space can be organised as the plant and materials supply point. Relative to the weekly construction programme, materials schedules for all trades, including all sub-contractors, can be prepared by the materials controller. In these circumstances, it will be found necessary to operate a material supply coding system (10) and to give thought to the order in which palleted materials are stored in the container (Monday morning's materials first). The container should be delivered to the appropriate floor level and the distribution at floor levels carried out by pallet truck, either hand or powered, as appropriate.

Pallet sizes and manufacture

The two most popular sizes of pallet are 1000 mm x 1000 mm and 1000 mm x 1200 mm, apart from these, specials are purpose made for use with particular loads. Pallets are usually of all timber construction, being of softwood boards secured to softwood bearers by nailing. In most cases they are of single bearer construction but when designed for very heavy loads sometimes double bearers are used. In this case, if pallet trucks are to be used, the ground clearance of the truck should be greater than the depth of the lower tier of bearers, otherwise the truck is inoperable.

Material waste

Categories of waste

1. Associated with the design process.

2. Factors particularly related to contract or administration and control.

Figure 7.1   A purchasing procedure for contractors operating under JCT Form

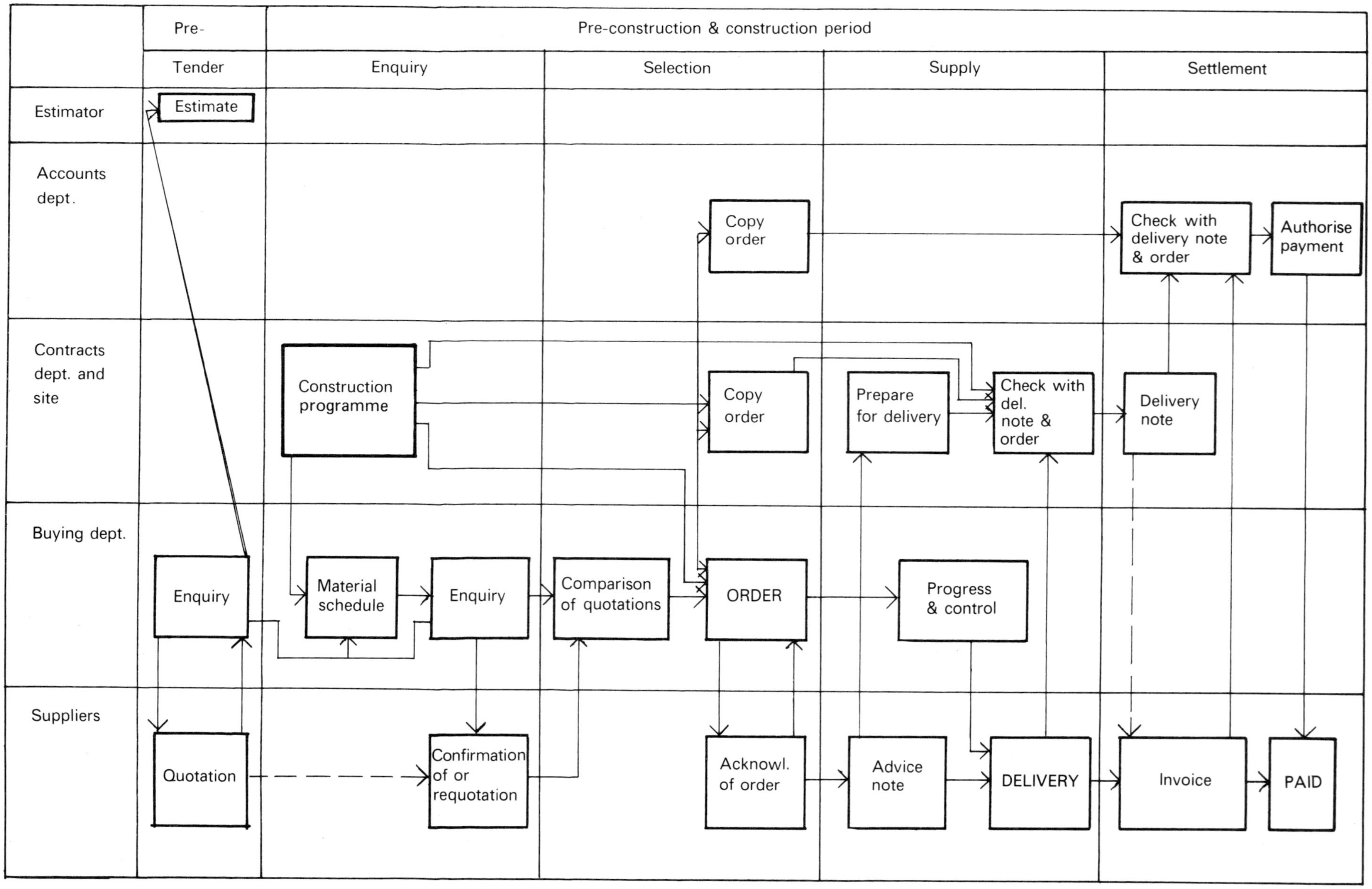

	Pre-tender	Pre-construction & construction period				
	Tender	Enquiry	Selection	Supply	Settlement	

3. Contractor using processes.

Associated with the design process

Any study of material waste must start with the design process. All too often the contractor's expertise is not used by the designer, due mainly to the design organisation being separated from production. Three questions which can be asked are:

1. Can the design be modified to reduce material loss in scrap?

2. Are the materials as specified, unnecessarily elaborated, ie is the manufacturer performing operations on his standard product which are unnecessary for the purpose to be used?

3. Is the designer over specifying? Do some structural materials have a fifty years life, whilst others have a two hundred years life span?

These three considerations are often outside the jurisdiction of the contractor. However, they are worthy of consideration by the speculative builder and may be applicable to the work of the builder doing repairs and alteration work.

Factors particularly related to contractor administration and control

Bad buying; this can be considered from a number of aspects:

1. Materials which are of an inconsistent nature.

2. Materials which are of too low a quality.

3. Materials which are of far too high a quality.

4. Over ordering, with resultant waste of the excess.

5. Under ordering, causing panic buying later.

6. Buying materials unprepared by the supplier, when clearly it would have been wiser to buy prepared and vice-versa.

7. Buying materials in random size when clearly precise lengths would have been cheaper and vice-versa.

8. Inaccurate taking off, both of quantity and description causing either shortage or excess, or the need to return unsuitable materials to the supplier and await the arrival of the specified material.

9. Poor programming and scheduling.

10. Poor administrative and communication procedures between the site, the head office, the supplier.

11. Lack of planning concerning handling methods.

12. Failure to gain a refund of VAT when eligible.

Contractor using process

1. Poor performance in the calling forward of materials.

2. Bad administrative and checking procedures.

3. Lack of planning concerning access, site roads, storage areas.

4. Failure to obtain protective sheeting, materials, etc.

5. Site plant unsuitable for conveyance of materials.

6. Materials ordered earlier than necessary.

7. Low level of security resulting in stealing.

8. Bad storekeeping and site requisition control.

9. Poor craft supervision resulting in high wastage by craftsmen.

10. Lack of planning in the cutting and fixing processes.

11. Poor level of site communications.

12. Little or inadequate consideration to protection of finished work.

13. Using materials for purposes for which they have not been specified.

14. Carrying out work not covered by the original contractual requirement or subsequent architect's instruction.

15. Bad bookkeeping: dayworks, etc.

16. When architect's instructions make materials in store surplus to requirements, failure to contact the architect to get his instructions for the disposal of the same.

17. Failure to gain refunds on returned cratage and drums, etc, when available.

18. Paying duty on fuel oil when duty free fuel was appropriate.

Figure 7.2   **A procedure for receiving materials on site**

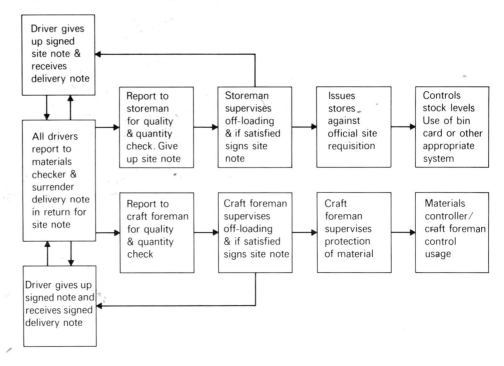

Some of the more important aspects related to communications

1. There must be a clearly defined purchasing procedure. Figure 7.1 outlines a procedure.

2. There must be a clearly defined procedure for receiving materials on site (see Figure 7.2).

3. There must be a clearly defined procedure for removing materials from site (see Figure 7.3).

4. When asking for quotations, it should be clear whether the supplier is expected to include in his quote for ancillary materials, cleaning, scaffolding, etc. Examples of these include:

(a) all rates for glazing shall include for leaving all glass clean both sides on completion.

(b) the sub-contractor is to clean up and remove from site all debris caused by his workmen.

(c) when sending out enquiries for ironmongery, it must be stated if the rates are to include for all necessary screws and fixing bolts.

Figure 7.3   **A procedure for removing materials from site**

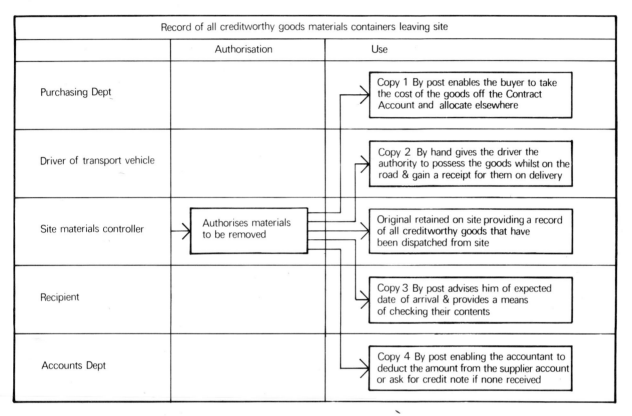

BIBLIOGRAPHY

1. Whitt, Dr. K.J., (Jan 1975), 'Where handling methods go wrong on building sites', Materials Handling News.
2. Rawcliffe, J.R. and Whitt, Dr. K.J., (July 1974), 'Getting the unit load right', Mechanical Handling.
3. Colechin, F.A., (22 Feb, 1974), 'Mechanising Housing Sites', Building.
4. Illingworth, J.R., BSc(Eng), (Dec 1974/Jan 1975), 'Materials Handling. Projected Developments. Building Technology and Management'.
5. Stratford, John, (Mar 1976), 'Materials Handling Plant', Civil Engineering.
6. Jolley, Derrick, (May 1977), 'Laing designs rig to lift cladding', Construction Plant & Equipment.
7. A Report, (June 1977), 'Roughing it on site', Materials Handling News.
8. Bourne, E. and Newbury, B., (23 July, 1976), 'Mechanical Plant Handling Equipment', Building.
9. A Report, (Dec 1973), 'More work on site handling called for', Clayworker.

10. Tatlow, David M., (Jan 1973), A paper delivered to National Materials Handling Centre Conference.
11. 'Material Movement', published by Massey-Ferguson.
12. Robey, I.C., (Nov 1970), Builders Merchants, Planned Deliveries, Building Technology and Management, IOB.
13. Telling, L., (Nov 1976), 'Materials handling and fork lifts', BTM.
14. Chandler, I.E., Materials Management on Building Sites (1978), The Construction Press.
15. Mudd, D.R., Suppliers invitation to tender for the supply of materials, IOB, Estimating Information Service No 19.
16. Tompkins, W.E., PPIOB, (May 1970), 'The purchasing power of a builder', BTM.
17. Dand, Richard and Farmer, David, 'Purchasing in the Construction Industry', Gower Press.

8. Manpower Planning

INTRODUCTION

This chapter is mainly concerned with people but it is probably valuable to examine first some facets of the construction industry before developing the main theme.

Official sources refer to the building industry and the civil engineering industry as the two constituent parts of the construction industry. It could be argued that there is a third equally important part, namely, the construction materials producing and extracting industry which not only supplies the materials but is also responsible for supplying considerable credit to construction firms to supplement working capital. The net output of these three partners in construction represents about one tenth of the gross domestic product of the United Kingdom. In the late 1970's, this combined production was around £10 000 000 000 per annum.

The civil engineering industry

Civil engineering is mainly concerned with the construction of roads, bridges, motorways, harbours, airports, power stations, and oil and gas installations. A large investment also takes place in sewerage and storm water schemes, flood barriers, reservoirs and water services. Nearly all of their products are high cost, long-term contracts, and only large firms have the essential resources to accept the role of main contractor on such projects. Some of the largest construction companies have literally scores of subsidiary companies with a total involvement in civil engineering, construction of buildings, materials manufacture and supply. Almost all civil engineering work is carried out by firms in the private sector, whilst the State is usually the sole client, in the form of local authorities, central government, nationalised industries and corporations. Therefore government restrictions on public spending usually act very directly on the civil engineering industry. At least, this was the case until the recent spending spree in the developing countries of the Middle East, Near East and some of the African states, which gave new opportunities to civil engineering and building contractors.

People in civil engineering

Site managers often hold degrees or higher national certificates in civil engineering. Their studies and experience are essentially specialist and the majority would not easily adapt to being a site manager on general building work and likewise the building site manager would not readily adapt to civil engineering work. These are different specialisms.

Civil engineering work, although heavily mechanised, is tough, often disruptive to family and social life and tends to attract the young and adventurous. However, many drift away to less demanding jobs during middle age, also during severe winters, which usually halt a large amount of civil engineering work, many leave the industry.

The Building Industry

The industry tends to be divided into sections although the section to which a firm belongs is not always clearly identifiable. There are many small firms concerned with maintenance repairs, and alterations and decoration of existing premises. Other firms concentrate exclusively on housebuilding activity for the private sector whilst others, under a heading of general building contracting, are involved with commercial and industrial, social, recreational and educational buildings.

Labour tends to be somewhat immobile between these sections so that a man employed on new housing work usually stays with this kind of work and does not become involved with maintenance and repair work or general building contracting.

Demand

Demand for building services has fluctuated strongly during the 1970's and, not unnaturally, has had a serious effect on the confidence of contractors to carry out forward planning and has also increased labour loss in the Industry as a whole.

"Five year forward estimates of public expenditure are prepared by the Government, usually in the winter of each year as a White Paper." (1). It is probably true to say that the greatest problem facing the construction industry is the need to stabilise demand. In recent years leaders of the industry have become extremely perturbed at the frequency and degree with which the Government has manipulated demand.

THE PERSONNEL FUNCTION

One of the primary concerns of the personnel service in any organisation is the promotion and development of industrial harmony. Associated with this a number of interrelated facets can be identified, namely:

1. Stabilising the labour force.

2. Wise pay policies.

3. Opportunity for personal development where appropriate.

4. Continuity of management style.

5. Providing a safe place of work.

6. Co-operating with trade unions.

Stabilising the labour force

In order to make recommendations on how the labour force might be stabilised it is necessary to find out why employees have left the firm and what the present employees think about their conditions.

The leaving interview	Irrespective of the reason(s) which cause a worker to leave an organisation, a member of management should arrange a leaving interview. In the case of head-office staff this is normally done by the personnel manager; in the case of the site worker, it may be the site manager for operatives with short service, but where a site worker has had long service with the firm it may be appropriate to invite him to head-office. Wherever the interview is held, the person organising it should make sure that the real reason(s) why the employee is leaving is known to management. A statistical analysis of the reasons why people leave the firm aids the personnel office when making recommendations for changes of policy.
Attitude surveys	Attitude surveys carried out by independent consultancies are now used extensively to find out how the firm measures up to the employee's expectancy. From the results a numerical score can be derived which enables a comparison to be made with previously conducted surveys in the company and a league rating in comparison with other firms, ie, interfirm comparison of employee expectancy.
Statistics	It is usual to measure the labour turnover and then analyse the figures obtained to attempt to find ways of reducing the wastage. There is a high cost in engaging and inducting any kind of labour.

$$\text{Separation rate} = \frac{\text{Number of separations during the year}}{\text{Average number employed during the year}} \times \frac{100}{1}$$

Reasons for leaving	There are a number of categories of separation, three of which are considered here:

1. Women who give up work on marriage or because they are expecting a baby, or the husband and senior breadwinner is taking up an appointment in a new district which necessitates moving home, or similar category of leaving.

2. Retirement.

3. Because the job has not fitted the person or the person has not fitted the job. This occurs not only with new starters but also with established workers when some change, real or imaginary, occurs.

Category 3 leavers should be sub-divided into those with less than one year's service, those with one to five, and hardcore leavers. The reasons for leaving should then be critically analysed. It is natural that there should be a higher incidence of leavers amongst short service employees than long, although it is still a financial loss area to the firm. From the findings of the leaving or termination interview, the personnel officer should assess the hypothetical question that, if given foreknowledge of the impending termination, would management have acted differently? The question should be asked, do resignations indicate a flaw in policies, practices or procedures? Do the figures point to a weakness in one department, manager or supervisor?

Site resignations	A number of different factors have to be taken into account when reviewing site resignations. An unskilled labourer may be approaching middle age and finding his work too physically demanding. Because of limited general or technical education or other reason of unsuitability, it may be impossible to progress him into another situation such as a materials checker, site timekeeper, or other similar less physical role. After a period of years of steady work in a district, building activity may be much reduced and an employee with five years service is now being asked to travel a considerable distance to work for the first time. Despite recogni-

tion by the firm of the new problem, he decides to make a change to reduce the amount of time spent travelling.

Apart from these normal reasons for severance, a relatively large part of the site population habitually does not settle down to long periods of stable employment. It is true that this is the manifestation of the problem, an Industry problem, not a firm problem. In the short-term, an individual personnel officer can do little to alleviate this.

Promotions.

The promotion of a member of staff sometimes unsettles others who feel that their own case for advancement is stronger than the successful person's. Underlying this is possibly poor promotion and advancement policies, an organisation structure problem, or a poor level of communication between the personnel officer or other managers and the aggrieved staff.

Management trainees and leavings

There is a particularly high cost factor in the training, recruitment and development of management trainees. The measurement of success can be carried out by the tabular approach shown in Table 8.1.

Table 8.1

1980 intake civil engineering trainees

Period of service	Number of separations	% leaving	% remaining
1980 - 81	10	20	80
81 - 82	8	16	64
82 - 83	6	12	52
83 - 84	4	8	44
84 - 85	3	6	38
85 - 86	2	4	34

The percentage leaving is expressed as a percentage of the whole. Ideally, it is expected to see a gradual drop in the number leaving each year as shown in the table.

It is also informative to analyse the length of service of the current labour force and to consider whether key posts are being monopolised by long-service workers (see Table 8.2).

Table 8.2

Length of service	Number of staff	% of total	High A	B	C	D	E	Low F
More than 15 years	6	6	4	1	1			
More than 10, less than 15	7	7	1	4	1	1		
" " 8 " " 10	8	8	1	1	4	2		
" " 6 " " 8	10	10	1	2	3	2	2	
More than 5	10	10	1		2	3	2	2
More than 4	12	12		2	2	2	3	3
More than 3	15	15				4	4	7
More than 2	15	15					3	12
More than 1	17	17					8	9

PAY POLICY

Modes of payment

A wage can be defined as a payment made to a worker in return for his labour. It is either expressed as a rate per hour, or per day, or per week.

| Flat rate | Some workers receive a flat rate, void of any additional allowances. |

| Standing bonus. | Sometimes a standing bonus of £x's per week is paid in addition to the flat rate. If a worker is absent for one day, he then forfeits one fifth of his bonus. Standing bonus has the same effect as a daily attendance bonus. |

| Piecework | A workman who undertakes to do work at a price per unit, for example, per metre super, is a pieceworker and generally speaking has given up most of his rights under the Employment Protection Act of 1975. |

| Financial incentive | The aim of an incentive scheme is to relate effort to reward. In October 1947 a joint agreement between the operatives' unions and the employers' federations was reached on the acceptance of incentive bonus schemes in the Industry. The basic concept was that an operative of average ability and capacity should have a reasonable opportunity of earning 20% more than the flat rate of wages by his own increased efforts. By the early 1950's, slightly less than half the building labour force had target bonus schemes in operation. Generally speaking, the smaller the firm the less easy it is to set up and operate a scheme. On repetitive construction such as new housing estates it is relatively simple to achieve a fair value for targets. In this case, the system of 'target operations' is appropriate. For example, a contractor and his operatives agree that the interior of a house can be decorated by two painters in one working week of forty hours. A choice is then made between the following arrangements (see Table 8.3. Assume a wage rate of £4 00 per hour). |

Table 8.3

Comparison between direct and geared incentive schemes
Anticipated input = 80 hours (2 men x 40 hours).
Planned level of bonus = 25%.

Direct schemes
Target time = 80 hours + 25% = 100 hours
All savings paid direct to operatives

Hours used	50	60	70	80	90
Hours saved	50	40	30	20	10
Total savings	£200	£160	£120	£80	£40
Payments each	£100	£80	£60	£40	£20

Geared schemes (50% gearing)
Target time = 80 hours + 50% = 120 hours
50% of savings paid to operatives

Hours used	50	60	70	80	90
Hours saved	70	60	50	40	30
Total savings	£280	£240	£200	£160	£120
Payments each	£70	£60	£50	£40	£30

If the input is less than anticipated the direct scheme favours the operatives; when the input is greater than anticipated the geared scheme favours the operatives. Use geared schemes to favour operatives on new untried targets.

| Wage negotiating body | The National Joint Council for the Building Industry is the responsible body for fixing the wage rates for the Building Industry; the rates are published in the National Working Rules for the Construction Industry. |

165

Composition of the wage

If the wage is made up of the constituent parts described in the Working Rule Agreement, the following would be considered:

1. Guaranteed weekly wage (the flat rate).

2. Guaranteed minimum bonus.

3. Joint Board supplement.

4. Incentive bonus payment.

5. Any overtime payments due.

6. Any payments due in respect of discomfort, inconvenience or risk.

7. Any payments due for carrying out work involving extra skill. These apply in particular to labourers acting as plant operators or vehicle drivers.

8. Tool money due to craftsmen who have available a kit of tools in accordance with a standard list. Reading some of the lists must arouse nostalgic memories in the minds of octogenarians. One cannot help but wonder how many bricklayers carry a pair of dividers.

The reality of the situation

In many areas of the United Kingdom the wages paid do not coincide with the rates quoted in the Working Rule Agreement and in fact, earnings fluctuate with the normal laws of supply and demand.

Administrative cost

There is a very high administrative cost in collating the information required to pay wages in the Building Industry and any moves which suggest simplifying the processing of wages should be looked at carefully.

Fluctuations in take home pay

One of the most common causes of complaint by the labour force is the fluctuations in take home pay which take place due to one or a combination of the following factors:

1. Changes in demand, nationally.

2. Changes in demand, locally.

3. Inclement and seasonal weather.

4. Lack of balance in incentive targets.

5. Often a good deal of discretion concerning bonus payments is allowed at site level so that if there is a local scarcity of men in a particular craft, payments to this craft are increased, creating a lack of uniformity in the wages of the different crafts.

6. On a badly managed project, site management may move from crisis to crisis. On such sites, wage rates may be subject to frequent manipulation in an attempt to cope with the most recent crisis. Instability of wage levels can only have an unsettling effect in the long-term.

Characteristics of a good pay policy for site operatives.

In the modern way of life, thrift is an almost forgotten word. Many families have a high proportion of their earnings committed to hire purchase payments of one kind or another. The central theme of any wages policy must be stability of earnings. A further consideration is that the earnings level of the various groups must bear relationship to the input required for that job. Closely associated with pay policy is the question of promotion policy. Operatives given the opportunity of further training and promotion, even though the promotion may not necessarily give an increase in weekly take home pay, but perhaps enhance such matters as pension or 'staff status', or holiday entitlement, will be inclined towards increased loyalty towards the firm.

Many firms offer some kind of inducement to long service at all levels of employment. For example a site labourer may be given paid holiday in excess of Working Rule Agreement conditions.

For some time now the Industry has operated with a relatively low basic wage enhanced by various types of bonus schemes and in many cases high overtime earnings. The resultant earnings level compares favourably with many other heavy industries. One might question the value of regular overtime in an industry which inevitably experiences difficulty in regulating the working conditions at the work face. To many, it just seems to be a device for paying ten hours wages for work which could be done in eight. Yet there are many firms which regularly work overtime. There is ample evidence to show that after prolonged overtime productivity falls, standards of safety fall and the quality of living is not improved. There seems to be a case of adopting a higher wages policy without necessarily increasing earnings. Overtime working often involves site managers staying on site longer. Many site managers do not receive extra pay for overtime hours so that it simply dilutes their pay.

Managers' salaries.

The Diamond Commission's report 'Higher Income from Employment' made comparisons between the salaries of British managers and the shop floor and also between British managers and managers in other European countries. In both of these exercises, British managers faired badly. Differentials between managers and many of the shop floor workers have been eroded.

On quite a number of sites workmen have greater take home pay than the site manager. Although this may be offset to some extent by greater security and fringe benefits, the harsh fact is that many managers are not adequately rewarded for the responsibility which they carry. Also, for many, site management is a terminal role and the significance of this is important. It is possible to pay a manager less than a workman without destroying the manager's motivation when the manager is on the way up the ladder of success and will ultimately earn more than the workman, When the manager is in a terminal role and exposed to a condition of under-payment compared to the workman, he must inevitably become cynical and critical of his employer.

Military history contains more than one reference to castrophe arising from underpaid infantry officers and soldiers. Site managers are in charge of the front line troops and can be regarded as a construction company's most prized asset, even if not figured on the balance sheet.

Site managers with potential could reasonably expect a career pattern where, through a mixture of training and experience, they pass up a grading system which is accepted by the whole of the Construction Industry.

TRAINING AND DEVELOPMENT

Training personnel is an important part of the work of the personnel officer. The company must establish first what its training policy is, what it hopes to achieve from training and how much it proposes to spend in the next five years or so on training. The Industrial Training Act of 1964 removed the voluntary concept and all except the very smallest firms had to pay a levy to The Construction Industry Training Board, one of twenty-eight industrial training boards set up under the Act.

In order that smaller firms could have the benefits of a training officer, many now group together for training purposes, so that the group employ the training officer and his costs are recoupable by the group from the CITB under the grant system.

Training policy

In order to develop a suitable training programme, a firm must analyse its needs in the light of the present and future work load and the present manpower position and the manpower requirements for the future. From this it should be possible to develop an order of priorities, and a cost and time budget for carrying out the training. Many firms appoint a director to have overall responsibility for training, whose role is to recommend training policies to the Board and see that they are carried out.

Levels of training

1. Semi-skilled workmen

2. Craft apprentices

Training in profession skills

3. Site management

4. Site surveyors

5. Estimators

6. Buyers and others.

Training for higher management will often be concerned with meeting requirements highlighted by the firm's succession plan.

Courses and Education

The Construction Industry Training Board have a large training centre at Bircham Newton in Norfolk. An important and very successful venture at Bircham Newton has been the plant operators' course. The rural location of the centre makes it ideal for this purpose. A major part of the work of the CITB has been to develop training where it was clear there existed a need but no provision had been made by existing training establishments. They have either set up courses themselves or assisted others to arrange them in the following skills:

(a) civil engineering apprentices

(b) scaffolders

(c) barbenders

(d) formworkers

(e) plant operators

(f) general foremen

Craft apprentices

Craft apprenticeship courses were established throughout the United Kingdom in the immediate post war (1939-45) era. At that time a five or six year apprenticeship period was usual, varying by one year according to the particular craft. 'On the job' training was supplemented by attendance for one day per week at the local technical college until the apprentice was eighteen years of age. Since that time a number of studies and reports have stated that a more effective method is by full-time periods of training. A boy leaving secondary school now enters into an apprenticeship agreement with a building firm and commences immediately a full-time training programme, usually at the local technical college. This period of training is typically twenty-eight weeks. After this follows a two and a half or, in some cases, three years' apprenticeship. During this period, the apprentice attends technical college one day per week. Every effort is being made to fashion the training relevant to the needs of industry.

Site managers.

Before entering into a management traineeship with a construction firm, many potential managers have a full-time period of building education in a technical college. Formerly, this often consisted of two years studying for an Ordinary National Diploma in Building followed by two years further study for a Higher National Diploma. If the Higher National Diploma was done on a sandwich basis, it would take three years, since two six month periods would be spent on site with an approved employer. However, as a result of the Haselgrave Report (2), these courses may now be phased out and replaced by Technician Education Council Courses. Those in employment take Technician Eduction Council (TEC) Certificate Courses, whilst full-time students attend TEC Diploma Courses.

The former idea of having an end of course examination as the deciding factor of whether a student progresses to the next year is gone. Indeed, the courses are not divided into years but based around subject modules of sixty hours' duration each. Some modules require that a student should have completed certain other modules before attempting them. Replacing the former end of session 'make or break' examination is a system of continuous assessment. This system makes a serious attempt at measuring the students' input throughout the whole of the course, and demands a higher standard of professionalism from college lecturers than the old method of assessment.

Management traineeship

Most large contracting firms now operate well planned training schemes. These are usually a combination of induction to the company by working for a short period of time in two or three departments intermixed with in-company training modules. Subjects which are emphasised currently are industrial relations and safety on the site. A major objective of training is to accelerate the rate at which the trainee matures so that he can fulfil a productive role earlier in his career.

Qualifications for managers

The job descriptions in the Appendix indicate the qualifications appropriate to the various roles. It is now abundantly clear that the premier qualification for all contractors' management staff is full membership of the Institute of Building.

Continuing Education

In the early nineteen seventies, discussions took place between the National Federation of Building Trades Employers and the Institute of Building concerning the large number (at that time estimated to be one hundred thousand) of site managers who had never received any formal training.

As a result of these discussions and following preparatory work, a scheme of continuing education was developed, the major criteria for acceptance on the Institute of Building Site Managers' course being that a person must be more than twenty-five years of age and have had more than two years experience beyond craft foremanship. These courses are now in existence at approximately twenty colleges in England. This is a unique experiment to satisfy a particular need but because of the rapid rate of change in the development of technology, continuing education is likely to become common for all grades of management in the future.

Planning Training

In order to measure training requirements, top management must make a diagnosis of the current state of the firm, consider its effectiveness now, forecast where the firm should be in the next five years and what manpower will be required to achieve this aim. There are a number of good reasons for having a management appraisal system. One could argue that without an appraisal system, it is very difficult to provide effective training.

When appraising each member of the management team they will look at the individual's past performance and at his potential contribution in the future. In making the assessment, they must take into account the career aspirations of the individual. After this initial assessment, hopefully, the personnel manager is able to judge what type of appointment each individual is likely to be most successful at.

When the whole staff has been appraised, training programmes should be developed which lead towards satisfying the future needs of the firm and also meeting the reasonable needs of the individual.

Table 8.4 **A typical performance appraisal form**

Confidential

Site manager . . . . Contracts manager . . . .
. . . . Appraisal period ending . . . .

Negative	−3 −2 −1 0 +1 +2 +3	Positive
A Inflexible, lacks confidence to instigate change from original plan.		Responds quickly and efficiently to changing circumstances.
B Delayed, incomplete, or ineffective administrative work.		Prompt and thorough treatment of administrative work.
C Intolerant, autocratic. 'Tells' leadership style.		Develops and maintains good relationship with his support team.
D Lacks bearing, posture, presence.		Presents a good company image towards the client, architect, clerk of works.
E Follows the firm's established routines only, lacks vision.		Good innovator, constantly developing new practical ways of doing work.
F Does not recognise a developing industrial relations problem until he comes face to face with it.		Sensitive to the feelings of the operatives. Gets on well with trade union representatives.
G Insensitive to the need for safe, clean and tidy work conditions.		Has a real concern about safety, health and welfare of operatives.
H Little or no interest in developing subordinates 'Task' oriented.		Very good at developing subordinates and talent spotting.

| **Performance appraisal is an ongoing process** | Once a performance appraisal system has been launched, it is a continuous part of the management development system. |

Superiors will be expected to appraise the performance of subordinates at regular intervals in order to measure progress against the initial assessment. Ideally, more than one superior should assess the performance of each subordinate to avoid bias arising out of personality clashes. A typical performance appraisal form is depicted in Table 8.4.

Minus three and positive three represent two extremely different descriptions of behaviour. The intermediate ratings represent degrees of behaviour between these extremes. The contracts manager would be asked to base his ratings solely on his own observations based on the appraisal period just completed. Appraisal periods are often of six months duration. At the end of each period and after completion of the performance appraisal, a review interview should be arranged. The person completing the form should not complete any section unless he has had ample opportunity to observe the appraised person in that particular characteristic.

Review interviews

It can be said categorically that if review interviews become an ordeal for those taking part or if salaries are discussed, the interview system has already failed or is well on the way to failure.

Performance interviews should permit a frank and friendly exchange of views and should be directed positively towards developing the individual and satisfying the firm's manpower requirements. If a manager looks forward to his appraisal interview and prepares for it, if in his mind it becomes synonymous with sound management policy, then it is possible that the interviews are being conducted along the right lines.

Judge, jury and prosecutor

The success of a management development system hinges very critically on the kind of atmosphere which is created at the interview. It must be an atmosphere of partnership in reviewing the past performance and seeking new goals. It would be well to have a large placard on the wall, with the statement *'Those people who don't make mistakes don't do anything'.* Every effort should be made to avoid damaging the self-esteem of the manager. Before raising any point, the interviewer should mentally consider the following three questions:

1. Is it true?

2. Is it fair?

3. Will it result in an improvement in performance?

The atmosphere created should be one which encourages the person voluntarily to want to talk about his shortcomings and how performance might be improved. It is folly to rush appraisal interviews and the interviewer should have a good ability to listen. The first interview with any manager may not be very fruitful. Interviewers should be patient and regard the first one as setting the scene for subsequent fruitful interviews. Any interviewer who sets himself up as judge, jury and prosecutor will get the results he deserves.

Personal assessment

Some organisations ask the manager to complete an assessment sheet of his performance before attending the interview. Table 8.5 shows a suggested layout.

Table 8.5		**Personal appraisal form**		**Name**			

Productivity -3 -2 -1 0 $+1$ $+2$ $+3$

Is the contract being run to programme?
How effective have you been in controlling progress?
Comments (be as specific as possible).

Quality

To what degree are you effectively combining effort and time to produce good quality work?
How do you rate your success in controlling waste?
Comments (be as brief as possible).

Efficiency

How efficient have you been in controlling plant and labour?
Comments.

Self improvement

Looking back on the in-company course on . . . . attended in . . . . How much benefit do you think this course has been to you on the present contract?
How much benefit do you think you have derived from the last appraisal interview?
Comments.

The danger of self appraisal forms

Forms such as the one shown in Table 8.5 are just as likely to put the manager into a self defensive position as they are to do any good. Their use is therefore questionable and if used, the wording of the form is very important.

MANAGEMENT SUCCESSION

As men approach retirement, it is obviously necessary to consider who will fill the vacancy created by the retirement before it occurs. Also, in the event of a vacancy occurring at senior level for unpredictable reasons, there should be someone earmarked for the position before the vacancy occurs, or it should have been decided before the vacancy occurs, that when vacant, it is to be filled by an outside appointment. If a manager begins to show signs of strain in the fifties age group and the particular role which he fulfils is a high pressure role, a wise management will give consideration to a pre-retirement lateral transfer, if such a transfer would reduce the pressure on the individual. Fifty-seven years of age is a statistical peak for men concerning many of the typical physical problems, strokes, heart conditions etc.

The managing director will take complete control of the succession plan for senior managers. Whilst he would obviously listen to advice.

from the personnel manager, it must be the managing director's decision A management succession plan should be a confidential document. Men who have spent considerable time in head office posts, near to the hub of things, are usually fairly flexible and can adapt to two or three posts. Men who have spent a lifetime on site tend, in the main, to be less adaptable and when such men, through reasons of ill health, need to be moved to a less demanding role, difficulties can arise. Site management is a terminal role for many site staff. Figure 8.1 shows a succession plan.

Figure 8.1   **Pyramid of functions from management traineeship to site manager role**

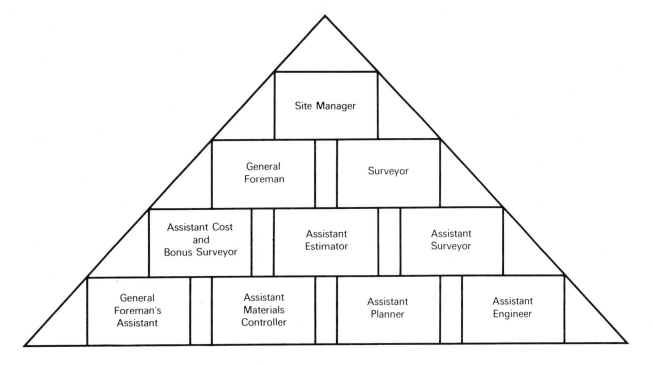

OPPORTUNITY FOR PERSONAL DEVELOPMENT

To consider motivation as if some people are highly motivated from birth and others have low motivation is a nonsense. The great influencing factor is the kind of environment to which the person has been exposed or is currently exposed. There is a wealth of meaning in the two sayings, "born with a silver spoon in his mouth", or "hungry boxers become future champions". The role of the personnel officer is to detect where hidden talent lies and provide the opportunity for developing it. To detect where frustration exists and providing its removal accords with the firm's goals, to remove it. Every manager and supervisor is involved in the personnel management function. It is the duty of each one of them to know the vocational aspirations of each member of their term and where these do not clash with the firm's objectives, to help them to attain satisfaction.

"The manager who, in response to ideas current in today's society, slavishly adopts a participative style of leadership, may run almost as great a risk of being out of tune with his staff's wishes as does the manager who persists with the old authoritarian style of leadership. No one style is likely to be appropriate in all situations, or to be the most suitable general style for all work groups. Whatever style a leader adopts, if he does so

automatically and unthinkingly, he risks failing in his task as a leader. Successful leadership involves being aware of the requirements of each situation and using the appropriate skills to deal with it. It does not demand that the leader adopt, in all situations, a set of techniques prescribed by some formula."

Roger Hughes and Vivian Shackleton (3)

MANAGEMENT STYLE

The most appropriate management style for a firm is not a static factor. In periods of growth or recession, or periods of readjustment, the management style is a matter for serious consideration. In the present post-Bullock era and the promise of legislation concerning worker participation in management (Prime Minister Callaghan, May 1978), firms should be examining their leadership style and assessing its appropriateness for the future.

On the construction site and during the actual construction processes, it seems desirable that authority must be vested in management through one person for safety reasons. It is difficult to envisage a situation where the worker had a primary right to negotiate on the method and organisation of work. Yet this applies in some European countries, notably Yugoslavia and closely followed by Sweden. R Tannenbaum & W H Schmidt (4) discuss four styles of leadership and these are referred to in an interesting publication by P J Sadler (5). These are as follows:

1. Tells.

2. Sells.

3. Consults.

4. Joins.

These four styles represent the full pendulum of management style. 'Tells' reflects the manager who makes his own decisions and then tells his team what to do. The 'Joins' style involves delegating to the group the right to make decisions.

The effectiveness of a 'Joins' style of management would be influenced by the current atmosphere between management and trade unions or indeed between union and union. The Germans possibly have an advantage in this respect over Britain in having one union for each industry. Thus any differences between factions on the union side are not institutionalised. However, should there be only one style of management for a firm or does site work lend itself to a different style of management compared to the head office? Would a 'design and build' site operation warrant a different leadership style than a competitive tender contract?

These may appear frivolous questions but an industry just moving away from 'hire and fire' practice and the 'lump' system reveals remarkable differences in management style between the best and worst managed firms. Earlier in this chapter, mention was made of construction firms employing consultants to conduct attitude surveys amongst their employees. Valuable information can be gathered by this means on the reaction of workers to the management style.

Difficulties in implementing a 'Joins' style of management

Quite frequently a high percentage of a contract is sub-contracted and the duration of the stay on site of many sub-contractors may be short.

174

For obvious reasons, such men might not wish to become involved with participative management on any one site. If the worker representatives are elected from the main contractor's employees, there is a fair chance that they will be tower crane drivers, fork lift truck drivers, storekeepers, or fulfilling some other key role. One might question who is going to foot the bill for the increased cost of decision making, assuming that it is possible to have representatives from the work force without thoroughly disrupting production. If they are all elected from the main contractor's employees for the reasons mentioned earlier, the increased job satisfaction derived from this by the sub-contractor's employees may not be considerable. What is an appropriate style of leadership for a firm requires analysis of the situation, selection of the appropriate style; then follows the instruction of all managers in the selected style.

PROVIDING A SAFE PLACE OF WORK

It is essential that site managers are properly trained in safe constructional practices and also that they are well briefed on the legal requirements concerning safety. It comes within the personnel manager's role to arrange that managers should attend appropriate courses either in-company or external. It will also be necessary to arrange training courses for first aid personnel, plant operators, timbermen, and others whose negligence or ignorance might risk lives. When the firm has incurred cost in training personnel who are on the same terms of employment with the company as the ordinary workmen, the personnel manager may wish to recommend to his superiors that these key workers enjoy staff status, have a higher degree of security than the ordinary workman, and are required to serve a longer than normal period of notice when leaving, to provide a measure of protection to the firm against the sudden loss of key workers.

INDUSTRIAL RELATIONS

"We think it important that positive employment policies should be introduced not simply by decree on the part of management but that relevant policies should be the subject of genuine consultation and, where appropriate, negotiation with the workers through their union representatives."

Phelps-Brown Report (5)

Trade Unions

The two strongest unions in the construction industry are UCATT, the Union of Construction Allied Trades and Technicians, and TGWU, Transport and General Workers Union. The Trade Union and Labour Relations Act 1974 did not repeal the Code of Practice associated with The Industrial Relations Act 1971. This Code, accepted by the two major political parties, stresses the value to industry, as a whole, of strong representative trade unions.

Whilst the building industry does not have an unusually high incidence of disputes or strikes, considerable improvement could be made in the industrial climate. However, unlike manufacturing industry which has a permanent location, most building projects are of a relatively short duration and thus do not warrant separate negotiations concerning the application of trade union machinery on site. Yet there is a uniqueness about most building sites.

Trade unions membership tends to be strongest in the major cities and here the trade union members generally expect to have a proper representative system of negotiating machinery with site management. However, in many small towns and rural areas membership is weak, and there could be difficulty in setting up a trade union negotiating body on site. Very often in a smaller firm there is a good deal of flexibility in the approach to doing work. If at any time a dispute arises, this flexible approach creates a problem, since an arbitrator must look for an accepted standard, namely, the Working Rule Agreement.

The Working Rule Agreement, regarded by many as being inflexible, provides a basis from which to commence negotiations and it also highlights areas of difficulty where men may require extra remuneration.

Mention was made earlier of having a 'Joins' style of management on site. When setting up trade union negotiating machinery on site, some of these same problems may reappear and also new ones as well. If much of the work is being carried out by sub-contracted labour, then site management may find itself involved with too many different trade unions and there may be difficulty, on the trade union side, in identifying common objectives. Where trade unionists are in a minority, a trade union representative may be unacceptable to the non-unionists, raising a further aspect of split representation for the work force.

The site manager role in industrial relations.

It is essential that site managers are well briefed in current employment legislation. Since 1975 there has been continuing change in employment legislation. It is not within the scope of this book to delve into the details of this legislation, but it is clear that, for a person embarking into site management, employment law may require greater consideration than construction law. Hence, site managers need an initial course of training in employment law and then constant updating. Many firms use the courses organised by the Building Advisory Service of the National Federation of Building Trades Employers.

Apart from familiarity with the law, the site manager needs instruction in how to conduct joint consultation with his operatives In-company training schemes based around role playing exercises are extremely popular. The end-product of these preparations is to issue the site manager with a manual of procedures which helps to standardise the firm's approach to joint consultation with the trade unions. The manual should help the manager to understand in what circumstances he should act without reference to head office. In other circumstances, he will be advised to act but report his actions to personnel, or, in large organisations, the specialist industrial relations officer. The third set of circumstances described would inform him to take no action but report to head office and await instructions. This approach gives him freedom to act in the routine, but as circumstances reach crisis point, the need for the intervention of an industrial relations specialist arises.

Summary - Industrial Relations

"At site level management seems often faced by a situation of constant crises and of pressure of deadlines in which the orderly planning of labour requirements seems scarcely practicable ... Throughout our in-

vestigations, we have seen repeated evidence of the lack of management skills in those closest to the point of application, ie, the supervisor."

Phelps-Brown Report, paragraph 291

The extract quoted shows the vital link between good management of the production processes and industrial relations. When men find their work interrupted through lack of information or resource failure, or when work must be destroyed through error, whatever the cause from which the error arises, and when waste occurs due to poor communications, or lack of co-operation, then morale sags and as the frustration mounts and dissatisfaction grows, some men leave the organisation, others become resigned to a hangdog attitude whilst others look for trouble. These are not the attitudes which management wants to create.

Attitudes which management is trying to foster.

1. The satisfaction which comes from pride in the completion of a good job.
2. The satisfaction which comes from being a member of a well respected team.
3. The feeling of pleasure which arises from positive physical and mental effort.
4. Pride in the leadership.

Essential desires of men at work

1. Freedom from the fear of dismissal and unemployment.
2. Freedom from racial discrimination.
3. A high standard of net wages.
4. An orderly and safe workplace.
5. Wise leadership.
6. Reasonable hours of work.

BIBLIOGRAPHY

1 The Public Client and the Construction Industries. HMSO.

2 Report of the Haslegrave Committee (1969).

3 Hughes, R and Shackleton, V (11 September 1975). Leadership styles. New behaviour.

4 Tannenbaum, R and Schmidt, W H. How to choose a leadership pattern. Harvard Business Review, Vol 36, No 2.

5 Sadler, P J. Leadership style, confidence in management and job satisfaction. Ashridge Management College.

6 Report of the Committee of Inquiry under Professor E H Phelps-Brown into certain matters concerning labour in building and civil engineering (1969).

9. Plant

Introduction

It would be quite wrong to contemplate writing a book on building production without at least one chapter concerned solely with plant. To attempt a complete treatise on the subject, both from technical and managerial aspects, would cover many volumes and require many experts to complete the task. The rate of development in plant design and usage is such that inevitably, at least in technical content, the treatise would have in-built obsolescence before publication. It is therefore wise to set rather precise parameters for this chapter at the outset. Consideration is given here only to the sort of plant which might be owned or directly hired-in by small or medium sized contractors. Furthermore, no attempt is made to delve into the technical aspects of plant usage and its main purpose is limited mainly to matters appertaining to plant administration, organisation and control which come within the orbit of the contracts manager and the site manager.

 A plant department may be involved in providing the following services to sites:

1. Hoardings, gates, fencing materials.
2. Temporary roads, bridge materials, vehicle wash down plant and equipment.
3. Site area lighting equipment.
4. Temporary buildings, furniture, canteen equipment and the necessary lighting and heating, toilets and washrooms.
5. Excavating equipment and associated plant, road transport vehicles to dispose of the spoil.
6. Materials handling plant equipment.
7. Plant and equipment for concreting work.
8. Scaffolding for both internal and external use.
9. Protection materials and equipment.
10. Formwork of timber, steel decking, adjustable propping equipment.
11. Workshop fitters and welders to maintain, modify and repair equipment and plant.

The reasons for the rapidly increased usage of mechanical plant

Spin-off from other industries

Some industries have passed through the mechanisation phase and are now well advanced in the automation phase. The construction industry is enjoying a spin-off from the research and development work of other industries. For example, fork lift trucks have been in wide use for more than a quarter of a century in factory warehouse situations but only in recent years have they become common on building sites. It is perhaps hard to understand why it has taken so many years for the industry to take advantage of this mechanisation. One aspect which should not be ignored is that, in a situation where perhaps eighty percent of the production process is subcontracted, the difficulties of initiating a system of materials ordering, supplying and handling, etc, which would be acceptable to all on one site, are immense. Often the main contractor lacks incentive to set about this co-ordinating role because of the lack of planning time and the fear that he may incur costs which are not in any way recoverable from some of the sub-contractors. For example, a plastering sub-contractor may normally expect to have his materials delivered to each floor level by the main contractor's hoist and then arrange his own distribution at that floor level. The introduction of a fork lift truck enables the main contractor to supply the materials to the balcony of each flat so that the plasterer's labourers' movement time is reduced by seventy-five percent. In times of labour shortage in the plastering trades it is questionable whether the unit rates for plastering would be reduced pro rata as a contribution to the main contractor's costs.

Changes in constructional form

The post-war period is noteworthy for the changes from a mainly craft-based industry to forms of engineering construction. These new forms create the need for a much higher plant input.

Social acceptability

Many traditional manual building operations are both fatiguing and messy. Within this category are included such tasks as:

(a) hand mixing of concrete

(b) hand digging of trenches, etc

(c) wheelbarrowing bricks, blocks, aggregates, etc.

One of the reasons why these manual processes are now less acceptable to workmen is because some industries have almost completely eliminated hard and dirty jobs. The more congenial the conditions of employment, the more people are likely to make a life career in the industry as opposed to the drift away from the construction industry at early middle age.

International comparisons concerning productivity

There have been a number of surveys in recent years demonstrating that the output of the British worker is lower than his counterpart in various other countries. However, productivity is a ratio between input and output. Input includes the amount of plant that a worker has at his command. In this respect, mechanisation obviously reduces the labour required and increases the plant/worker ratio. The British worker has less plant at his disposal than his colleagues in other leading western countries.

The 'time' dimension

Clients invariably require the shortest construction time period compatible with a well-constructed building. The client wants early occupation of the building in order to conduct his business. Capital invested in a partially completed building gives no return. The builder presenting the shortest time production programme at the negotiating stage often wins the contract.

Turnover

A contractor presenting the lowest tender usually receives the contract. For this to be viable, there must be high turnover related to capital involved.

With the ever increasing complexity of legislation and industrial relations, the overheads to turnover ratio moves in an adverse direction. Speedy construction is therefore desirable. Mechanised operations as opposed to labour intensive usually present the best opportunity for attaining these ends.

Land usage and building design

High land values, particularly in urban development, necessitate the consideration of deep basements with the usual attendant excavation problems and high rise construction with associated lifting and component assembly problems. All this creates the need for new and better plant and equipment.

Setting up a plant department in a small contracting firm

Before setting up a plant department the management need to give serious thought to the reasons which cause them to believe that it is a worthwhile action to take. On the credit side, they probably hope for a reduced cost in plant operations and a greater degree of control over plant resources. On the debit side, they are planning to employ capital in plant and money which is not easily recovered should the venture prove to be unwise. In using capital they are reducing the amount of liquid assets, which may in the short term reduce the total value of work which can be safely undertaken per annum. It can be argued that because they have invested more money in fixed assets the firm's credit worthiness is improved, but against this it can be said that the greater the reliance on borrowed capital, the more a firm is at risk. It is essential, therefore, for the management to establish policy guide lines. It is likely that they would need to prepare a timescale for the development of the department. Coupled with this would be the release or obtaining of finance to pay for the development.

Having decided what services should be provided, the policy concerning internal plant hire rates would have to be discussed and developed. Unless a full budgetary control system is introduced, there is a possibility of providing a loss-making service without even realising it. Ideally, charges will be levied against the plant department for establishment and service costs in which they participate. These would include rents, rates, lighting, etc, related to the head office and such matters as a share of the costs of personnel, wages, canteen, etc. Apart from these items, there are obvious overhead costs related to the running of the plant yard. When all of these costs have been computated, hire rates for the separate pieces of plant can be calculated. Most firms would adopt the policy of providing a non-profit making service. Efficiency can be assessed by comparing internal rates with those prevailing in the market place. A direct absolute comparison between rates may be unfair if sites are demanding and getting an unusually flexible service.

In developing a policy for the plant department, the extent of the service provided may hinge around the following factors:

1. The availability of finance.

2. The quality of service already available from plant hire companies.

3. The potential usage of that section of the service.

4. The complexity of the machine or service in question.

Administration

Weekly costing system

It is important that a costing system is operated so that contracts can be levied with the true hire charges for plant. Costing also enables an assessment to be made of which operations are profitable and which are unprofitable. This gives guidance to estimators and planners on method selection and estimate rates. Costing systems also penalise the site manager who unreasonably keeps plant on site and this motivates him to carry out proper short-term planning.

Individual contracts are charged with the cost of the plant on site, excepting when plant breakdowns occur which are due to the poor condition of the plant and not related to inadequate site care and maintenance. Site managers should be in possession of target plant costs for the various operations so that they can assess their own performance. In the absence of a cost control system, it is difficult to measure the cost of delays for reasons such as bad weather, lack of information, sub-contractor failure to perform to programme and other factors beyond the control of the site manager. Great care needs to be exercised to bring the last stages of production to a speedy conclusion. There is, in many production processes, a not unnatural tendency to have a fall off in the productivity rate towards the end of the process. This sometimes leads to a loss of motivation, which is, of course, a different matter. It is the responsibility of the site manager to apply 'management by exception' principles at these times to ensure that plant is removed from site as early as practicable. Before scaffolding is dismantled or plant removed from site, there should be an inspection to ensure that the reasons for its presence on site have been fully identified.

Plant charges

Most mechanical plant which is delivered to site by the plant department will be charged on an hourly basis with a minimum hire period which is often one day. In this category would be included mechanical excavators, fork lift trucks and mobile cranage. If the plant is delivered to site by low loader, then there will be a delivery charge of £x per hour. If after completion of its work it is transferred to another site, the receiving site pays the delivery charge. If, however, it is returned to the plant yard directly from the first named site, then this site will be charged for both the outward and return journeys.

Some smaller pieces of mechanical plant are charged on the basis of a weekly hire rate. In this category would be included platform hoists, site dumpers, compressors and those pieces of equipment which do not require a specialist operator.

Small pieces of equipment such as locks, bolts, bars, lamps, ropes, etc, are charged and their full value added to the contract. If any of these small items are returned to the plant department, credit is then given to the contract for the marked down value. When a special piece of plant is purchased to carry out an operation peculiar to a particular contract and is sold on completion of the one contract, that site will be debited the loss between new and resale value.

Plant history cards

It is wise for the plant department to have an individual history card for each machine owned. The card is used to maintain a record of all work done by the mechanics, including replacement and overhauls and the cost in time expended. Such a record provides a basis for comparing the machines of one manufacturer with those of another. It can help to avoid undue waste in excess stocks of spares. Finally, it may help to decide when a machine should be sold or cannibalised for spares.

Depreciation

Records are maintained of the amounts charged to each year's trading in respect of the depreciation of plant and other assets. When these amounts have been assessed, money may be set aside in a 'plant sinking fund' so that when plant is declared unsuitable for use by the company, funds are available for the purchase of a replacement.

Table 9.1

Plant depreciation summary for the year ending 1st April 1984

Mixers	Value at commencement of period £	Value at end £	Depreciation £	Tax allowance £	Net depreciation £
MW 1	4 500	4 000	500		500
MW 2	8 000	7 200	800	400	400
MW 3	300	250	50		50
MW 4	6 500	5 800	700	250	450
Excavators					
ERB 1	20 000	18 000	2 000	1 000	1 000
ERB 2	18 000	16 200	1 800	800	1 000
Total	57 000	51 450	5 850	2 450	3 400

Tax allowance

In order to stimulate investment in plant and equipment and thus encourage industrialists to modernise, allowances to be set against taxable profit may be made in respect of monies invested in new plant and equipment. These are usually in the form of an initial allowance followed by a reducing allowance for some subsequent years. This is a very complex area and the province of the tax accountant. It is, however, worthy of mention that the wisest amount to depreciate a piece of plant may be related to the company profit for the year and not a standard way of calculation.

Methods of calculating depreciation

The straight line method

Assuming a contractor buys a piece of plant for £1 200 and anticipates a life of four years with a resale value of £200 at the end of the four year period. Table 9.2 shows the simple calculation to write off £1 000 in four years.

Table 9.2

Straight line depreciation

Year	Value £	Depreciation £	Reduced value £
0	1 200		1 200
1	1 200	250	950
2	950	250	700
3	700	250	450
4	450	250	200

Residual balance method

This method assumes that the greatest loss in value due to depreciation occurs in the early years of the life of a piece of plant, and if a contractor wishes to reflect the true value of his assets in the annual balance sheet, this should be taken into account when calculating depreciation. The assumption is made at the outset that a piece of plant loses value at a predetermined percentage each year, the same percentage being deducted from the annually reduced value.

Table 9.3 illustrates a piece of plant originally purchased for £1 200 reduced annually by 25%. It should be noted that the plant never assumes zero value.

Table 9.3   **Residual balance depreciation**

Year	Value £	Depreciation £	Reduced value £
1	1 200	300 (25%)	900
2	900	225	675
3	675	169	506
4	506	126	380

Summary of the two methods

If there is no intention of selling the plant in less than four years, it is arguable that there is little point in attempting to calculate precise intermediate year values. The sale price of a machine in four years' time will be influenced by market conditions as well as the state of the machine and hence it is impossible to predict with any degree of precision a future sales income. Thus, if tax allowance is not a consideration, the straight line method has much to commend it.

Actual sale

Although a decision may have been made at the time of purchase to dispose of a machine after four years of use, other counsels may prevail when the time comes. The following may apply:

1. The machine may still be in excellent working order.

2. There may be a cash famine which mitigates against all but the most urgent expenditure.

3. The organisation in question may be busy, but the industry depressed and thus the market conditions for selling plant poor.

4. There may have been no development in the four year period in that kind of plant and thus no opportunity to obtain a more up-to-date model.

Relationship between depreciation and costs

Anyone who has owned a motor car knows that with increasing age the annual costs of maintenance usually rise and in the life of many motor cars there comes a time when these costs can no longer be justified. The same applies to some items of contractor's plant. Generally speaking, it is a question of the cost relationship between the various parts of the machine and the durability of these parts. The parts, for many machines, can be considered as:

1. The engine and its ancillary mechanical components.

2. The electrical content.

3. Bushes and bearings.

4. The basic metal frame of the machine.

Some exceptions

It should be noted that it is not unusual for motorised road rollers to last for thirty years, whilst some of the better makes of site concrete mixers do last for between ten and twenty years.

Example 9.1

Optimum disposal time

Example 9.1, which is illustrated by Figure 9.1 and Tables 9.4, 9.5 and 9.6, demonstrates how one might decide in advance what the right time is to dispose of a machine. Naturally, machine records of past performance are necessary to arrive at the computation. Using the residual balance method, it is known that a certain machine can be written down by 25% of residual value each year. If these two, depreciating value and increasing maintenance cost, are graphed as demonstrated, the point of intersection is the right time to dispose of the machine. Assume the purchase price of the machine is £2 000.

Table 9.4

Depreciation by residual balance method

Year	Value £	Depreciation £	Reduced value £
0	2 000		2 000
1	2 000	500	1 500
2	1 500	375	1 125
3	1 125	281	844
4	844	211	633
5	633	158	475
6	475	119	356

Table 9.5

Maintenance costs

Year	Cost £	Cumulative cost £
0		
1	80	80
2	120	200
3	180	380
4	270	650
5	405	1 055
6	608	1 663
7	912	2 575

Table 9.6

Combined depreciation and maintenance costs

Line		Years	1	2	3	4	5	6
1	Depreciation	Singly (£)	500	375	281	211	158	119
2		Cumulative (£)	500	875	1156	1367	1525	1644
3	Maintenance	Singly (£)	80	120	180	270	405	608
4		Cumulative (£)	80	200	380	650	1055	1663
1 + 3		Total	580	495	461	481	563	727

The optimum time for disposing of the machine is after four years.

Inflation accounting

The inflation of the nineteen-seventies has brought into question the methods of allowing for depreciation just described. This method concentrates on the original value rather than the capital sum required to replace an item of plant.

Example 9.2

Traditional method of costing

	£
Capital cost of machine	10 000
Add interest on capital at 10% for five years' life	6 105
	16 105
Deduct anticipated trade-in value	2 000
	14 105

185

General maintenance, spares, fitters' time at 5% of capital cost per annum (5 years)	2 500
Major overhauls, two at 20% of capital cost	4 000
Ownership costs at 2% of capital cost per annum	1 000
Total outlay	21 605

Total number of working weeks in five years
    240 weeks = (5 x 48)

From records a utilisation of 75% = 180 weeks

Weekly hire rate £21 605 ÷ 180 weeks = £120.

NB: If the replacement plant costs £15 000, there has been a shortfall in depreciation allowance of £5 000.

Figure 9.1

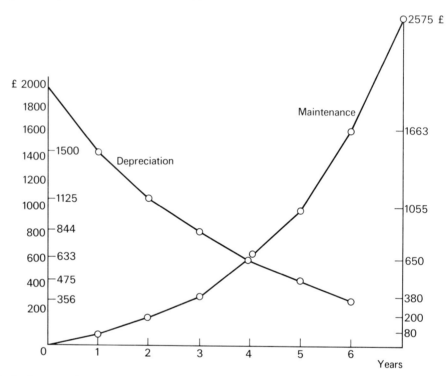

Example 9.3   **Inflation costing**

	£
Capital cost	10 000
Purchase price of replacement after five years, assuming 6½% inflation per annum	14 026
	24 026
Average capital cost	12 013
Add interest on average capital cost at 5% for five years	6 105
Gross total	18 118
Deduct trade-in value	2 000
Net total	16 118
General maintenance, spares, fitters' time, at 5% of average capital cost per annum for five years	3 005
Major overhauls, two, at 20% of average capital cost	4 810
Ownership costs at 2% of average capital cost	1 200
	25 133

186

From records 75% utilisation = 180 weeks

Weekly charge = £25 133 ÷ 180
= £140.

It will be noted that the operator's gross wages, fuel and insurance costs are not included in the calculation. It may be necessary to include a banksman's wages, depending upon the type of plant used. The weakness of the method described in Example 9.3 is that the weekly charge calculated would be high at the commencement of the five year period, correct in the middle period, and low at the end. Thus there is a case for adjusting the figures annually.

If the rate of inflation of plant cost is 10% per annum, then Table 9.3 adjusted for inflation is as shown in Table 9.7.

Table 9.7

Year	Value	Add 10% to value for inflation	Depreciation at 25% of inflated value	Reduced value
	£	£	£	£
1	1 200	1 320	330	990
2	990	1 089	272	817
3	817	899	225	674
4	674	741	185	556

Insurance

The contractor must take out various insurance policies. He must insure the Works in accordance with contractual conditions and he must also insure the workmen. He must also take out plant insurance, adequate to cover claims from the public at large, the workmen, and owners of property, and it is unwise to forget his ownership of the plant. It is perhaps worthwhile mentioning that insurance is like a common fund to which all contribute but only those who are unfortunate draw from. Further, those who draw from the fund may be asked to pay a higher than average contribution when next taking out insurance.

Plant insurance

Since the insurance companies are in effect making a charge to all contractor clients for managing a common fund, they have to have very precise agreements with their clients on what is being insured and what is not being insured. They want to know what plant will be operating on site, how it will operate, and where static plant will be located in order that they can properly assess the risk and from this the amount of premium to be charged. Thus an insurance policy which covers a tower crane in operation will not normally give cover whilst the crane is being erected and dismantled. A site dumper, covered adequately on site, may have inadequate cover whilst crossing the public highway.

Summary

It is important when negotiating insurance that the contracts manager knows how the contract is to be carried out and that the site manager is informed of what cover has been obtained. This information should be conveyed to the gangers and craft foremen who are responsible for getting the work done and all concerned should realise that if there is no necessity to make claims then insurance premiums will be reduced, resulting in greater profit, which is one of the main reasons for the company's existence.

Subsidiary plant company

Some construction companies form a subsidiary plant company in preference to having their own plant department. A plant company should be able to supply directly to the parent company a very high percentage of its total needs, hiring in from other plant hire companies only very special items, or for periods of exceptionally high demand. Such a subsidiary company normally has three primary policy objectives:

1. To supply all the plant needs of the parent company, either directly or by cross hire from other plant hire companies.

2. To make profit.

3. To distribute the risk of trading failure.

The parent company   In order to supply all the needs of the parent company, plant records must be maintained so that predictions can be made about future demand. If over a period of three years the demand for a certain type and capability of tower crane has been as shown in Figure 9.2 and the future demand is as shown in Figure 9.3, it would seem reasonable for the company to own three such cranes and hire-in for demand above this figure and hire-out when the company demand falls below this level. The plant company will obviously advertise its excess capacity through the plant hire contractors' association normal communication channels. In order to ensure that excess capacity is taken up by the market, the company may have to offer machines at below normal market rates. A plant hire firm may operate with two sets of plant hire rates, the rates charged to the parent company being marginally lower than those charged to other customers.

Figure 9.2   **Demand for tower cranes from historical records**

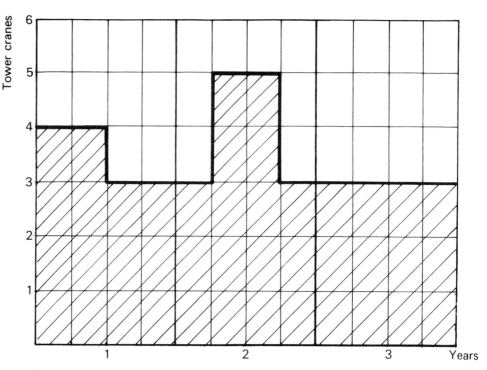

However, rates charged to outside hirers must be appropriate to those prevailing in the market place and the total income generated adequate to ensure plant replacement and the company at least not making a loss.

A plant hire company is not limited in size by the size of the parent company. It is usually able to provide a much more comprehensive service than a plant department. Being solely concerned with plant, the company has more market knowledge and expertise in plant hire than would normally be available in a plant department. Plant hire firms usually allow more favourable

Figure 9.3　　　　　　　**Forecasted demand for tower cranes**

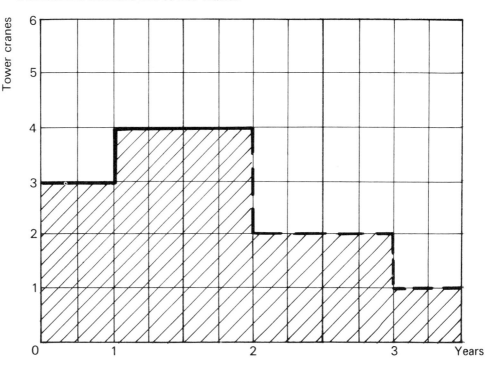

cross hire rates to each other than would be available to a plant department, although it will be appreciated that any big spender has muscle.

Complex machines　　In this context, a complex machine can be defined as one which requires a full-time specialist operator who would expect to be paid 'down' time. For many contractors, this is a luxury which he cannot afford, since, when the machine is not required, it is neither easy to sell the machine or dismiss the operator.

Hidden costs　　Where a machine requires a low loader to convey it from site to site, and if the department or firm does not own a low loader, they already have a dependency on an outside organisation. It would be unwise to purchase one unless a high utilisation is assured, since the low loader is a high cost machine and almost entirely unadaptable for purposes other than moving heavy equipment and plant from one place to another. This mitigates to some extent against owning plant which has to be moved from site to site by low loader which, if owned, may result in disproportionately high transport costs.

Plant with planned low utilisation　　A site manager may purchase a pump in order to keep excavations clear of seepage water. His most fervent wish is that he will not need to use it, but he must have it available as an insurance. Such plant should give many years of service if suitably maintained and protected. It may be planned that all concrete used on a site will be supplied by a ready mixed concrete organisation. At the same time, it may be wise to keep available a small concrete mixer, aggregates, and a small quantity of cement for the odd occasion when a half cubic metre of concrete is required. It would normally be uneconomical to obtain such a small quantity from a ready mix source. There are good arguments for owning both pieces of plant with their low utilisation factors, even though both must be available all the time but little used.

Simple machines with high utilisation　　This general category of company-owned machines contains those which do not require a specialist operator and can be categorised as simple. These include: site dumpers, platform hoists, compressors and site mixers. When these machines are not in use, the operators can be given other tasks to do.

Complex machines   When a contractor can clearly see a long run of work for a piece of plant which requires a specialist operator, there is obviously a case for purchase with the possibility of selling the machine when the long run of work draws to a close. The contractor must obviously have regard to finding suitable alternative work for the operator or negotiating his transfer with the machine.

Buying new plant   If adequate records are kept of all repairs and replacement parts fitted on company-owned machines, a judgement can be made on whether the ABC concrete mixer is a better buy than the XYZ machine. There are, of course, difficulties in making the judgement. The fact that two machines are the same age and have generated the same income from hire to sites does not indicate that they have been used the same amount or received equal treatment in maintenance and care. Comparison is fair only if the company plant returns show the hours when plant was in use, the hours when plant was on site but idle, and the hours when the plant was on site broken down. With regard to standardising maintenance practices, it is advisable that the plant department prepare routine maintenance schedules for the use of driver/operators. The site manager should arrange that time is available for maintenance and should check that the work is being done. This implies that all necessary greases, oils and cleaning aids are sent with the machine when delivered on site.

Making profit   Unless care is exercised, the plant hire company may become so enthusiastic about making profit from outside hirers that they provide a poor service to the parent company. If this unfortunate state of affairs occurs, it may be wise to allow the sites to hire-in from other hirers, if they believe this to be advantageous to them, until such time as the company's plant firm put their own house in order.

Summary   The devolution of a large company into a number of subsidiaries creates more specialists, often unleashes energy, enthusiasm and imagination. On the debit side, it usually increases communication problems and managers sometimes have different perceptions of the same reality.

Plant selection

If a highly satisfactory service is available from local plant hire firms for certain types of plant, it seems reasonable in the first instance to enjoy this service and concentrate efforts on owning those types of plant which are not well catered for by the plant hire firms. In judging what constitutes a satisfactory service, the following may be considered:

1. The competitiveness of the hire rates compared with those charged in other regions.
2. The availability of the machines required.
3. The age and physical condition of the machines available.

Discounts

Some firms offer better discounts than others, but the real questions are:

1. Which of the machines possesses the more reliable, robust engine?
2. Which of the firms normally provides the better after-sales service?
3. Which of the machines produces the best output?
4. Finally, which machine will be the safest in use?

PLANT ON SITE

Law relating to plant and vehicles

1. Factories Act 1961.
2. Construction (Working Places) Regulations 1966.
3. Construction (General Provisions) Regulations 1961.
4. Construction (Lifting Operations) Regulations 1961.
5. Health and Safety at Work etc Act 1974.
6. Control of Pollution Act 1974.
7. Highways Act 1971.
8. Hydrocarbon Oil (Customs and Excise) Act 1971.
9. Finance Act 1971, section 6.
10. The Hydrocarbon Oil Regulations 1973.
11. Petroleum Spirits Regulations.
12. Motor vehicles (Construction and Use) Regulations 1973.

The list provided is not intended to be comprehensive. The reader should be aware that law is not static, but constantly changing in the path of changing need. The site manager's main concern is to erect buildings and for most managers it is impossible to keep up to date on all legal matters concerning construction activity on site. It is necessary for at least one person in the plant section to be well briefed on legal requirements concerning the use of plant on site. This person should then give guidance to site managers; this guidance sometimes takes the form of a site manual.

Items one to four on the list are concerned mainly with the safety of men at work. They go into great detail concerning inspections and examinations of machinery and work places such as scaffolds and excavations. Table 9.8 provides the reader with most of the details.

The Health and Safety at Work etc Act 1974

This Act places responsibility for site safety squarely on both site management and site operatives. From this it follows that all employees should be trained in safe methods of working and good working practices. The Construction Industry Training Board, Bircham Newton Training Centre is fully equipped with all facilities and expert instructors, to provide training for almost all plant operators. Courses vary in duration according to the plant. The City and Guilds of London Institute have made significant strides in the development of courses for plant mechanics. It is the responsibility of the plant department to send only plant in good working order to sites. When plant is being transferred from one site to another, it is highly desirable that such plant is examined and tested by a competent mechanic so that the receiving site know that they have been sent a piece of plant in safe working order. It is a responsibility of site management to ensure that plant is not misused.

Control of Pollution Act 1974

One facet of this Act is that it gives local authorities the power to act when excessive noise is being made to the detriment of the public at large or the work force. Site management must not only consider the best location for a piece of static plant from a production viewpoint, but also consider locations away from noise sensitive areas. The use of sound baffles in the form of wooden screens or straw walling may have to be considered.

Electrically operated machines often present fewer problems from a pollution viewpoint. Such machines will usually be easier to start on a cold

Table 9.8 Guide to Statutory tests, examinations and inspections. Construction Operations.

Type of plant equipment or job involved	Testing and thorough examination	Who carries out this work	Results to be recorded on Form No.	Thorough examination	Who carries out this work	Results to be recorded on Form No.	Inspection to be carried out	Who carries out this work	Results to be recorded on Form No.	Legal reference
SCAFFOLDING							weekly or more often in bad weather	competent person	form 91 (pt I)	W.P. regn 22
EXCAVATIONS EARTHWORKS TRENCHES SHAFTS AND TUNNELS				weekly or more often if part has been affected, e.g. explosives collapse	competent person	form 91 (pt 1) entry to be made day of examination	at least every day or at start of shift	competent person		G.P. regn 9
MATERIALS OR TIMBER USED TO CONSTRUCT OR SUPPORT TRENCHES EXCAVATIONS COFFER DAMS CAISSONS							on each occasion before use	competent person		G.P. regn 10(1) G.P. regn 17(2)
COFFER DAMS CAISSONS				before men are employed therein and at least weekly or more often if explosives are used or any part is damaged	competent person	form 91 (pt 1)	daily and before men are employed therein	competent person		G.P. regn 18
DANGEROUS ATMOSPHERES				before men are employed therein and as frequently as necessary	competent person Instrument may be necessary	in any convenient way to show how examination was done				G.P. regn (21)(c)
CRANES (all types) CRABS WINCHES	once every four years and after substantial alteration or repair	competent person, normally by insurance co. engineer. manufacturer or erector	crane: form 96 crab: form 80 winch: form 80	at least every 14 months	competent person e.g. insurance co. engineer	form 91 (pt II) or on a special filing card containing the prescribed particulars	weekly	competent person e.g. crane driver	form 91 (pt I)	L.O. regn 10(1) (c) L.O. regn 28 (1) (2) and (3)
PULLEY BLOCKS GIN WHEELS SHEER LEGS	before first use and after alteration or substantial repair unless used only for loads under 1 ton	competent person, normally the manufacturer or insurance co. engineer	form 80	at least every 14 months	competent person, e.g. insurance co. engineer	form 91 (pt II) or on special filing card containing the prescribed particulars	weekly	competent person	form 91 (pt I)	L.O. regn 10 (1)(2) L.O. regn 28 (1) and (2)
HOISTS passenger	before first use, after re-erection, alterations in height of travel after repair or alterations	competent person, e.g. manufacturer, insurance co. engineer or erector	form 75 or form 91 (pt I) following alterations to height of travel	at least every 6 months	competent person e.g. insurance co. engineer	form 91 (pt II) or on special filing card containing the prescribed particulars	weekly	competent person, e.g. fitter	form 91 (pt I)	L.O. regn 46

Type of plant equipment or job involved	Testing and thorough examination	Who carries out this work	Results to be recorded on Form No.	Thorough examination	Who carries out this work	Results to be recorded on Form No.	Inspection to be carried out	Who carries out this work	Results to be recorded on Form No.	Legal reference
CRANES appliances for anchorage or ballasting				on each occasion before crane is erected	competent person, e.g. crane erector or fitter					L.O. regn 19 (3)
CRANES test of anchorage or ballasting	before crane is taken into use, i.e. after each erection or re-erection on a site or whenever anchorage or ballasting arrang. changed	competent person, normally crane erector in presence of insurance co. engineer	form 91 (pt I)	has to be done after exposure of crane to weather conditions likely to have affected its stability. A re-test might be necessary	competent person, e.g. insurance co.					L.O. regn 19 (4)
CRANES test of automatic safe load indicator (jib cranes)	after erection or installation of crane and before it is taken into use	crane erector or insurance co. engineer; must be a competent person with knowledge of the working arrangements of indicator	form 91 (pt I)				weekly	competent person, e.g. crane driver or fitter	form 91 (pt I) NOTE: this will be part of normal weekly inspection	L.O. regn 30
CRANES mobile jib test of automatic safe load indicator	before crane is taken into use, after it has been dismantled or after anything has been done which is likely to affect the proper operation of indicator, e.g. change in jib length	competent person, e.g. erector, manufacturer, engineer, insurance co.	form 91 (pt I)				weekly	competent person, e.g. crane driver	form 91 (pt I) NOTE: this will be part of normal weekly inspection	L.O. regn 30
LIFTING other APPLIANCES i.e. excavator dragline piling frame, aerial cableway or ropeway, overhead runway				at least every 14 months or after substantial alteration or repair	competent person	form 91 (pt II) or on a special filing card containing the prescribed particulars	weekly	competent person, e.g. driver	form 91 (pt I)	L.O. regn 28 L.O. regn 10
HOISTS (goods) made altered or repaired after 1st of March 1962	before first use and after substantial alteration or repair	competent person, manufacturer or insurance co. engineer	form 75	at least every 6 months	competent person, e.g. insurance co. engineer	form 91 (pt II) or on special filing card containing the prescribed particulars	weekly	competent person, e.g. fitter	form 91 (pt I)	L.O. regn 46

Reproduced by kind permission of George Wimpey and Co. The Table relates to Construction Operations only and not to Factories, Mines and Quarries.

morning, do not have 'dirty' exhausts, may be quieter running, and are also easier to switch on and off, so the machine may be shut down when not in use. Keeping machines well maintained and operating them with machine covers closed all help to reduce noise pollution. The local authorities are given power to act to prevent obnoxious gases from polluting the atmosphere.

Highways Act 1971, Clauses 31 and 32

These clauses apply when a contractor wishes to place a builder's skip upon the highway.

Items 8, 9, 10

These statutory instruments may have application when diesel fuel is being used on construction sites.

Diesel engines are much more common for contractors' plant than petrol. Petrol engines tend to be lighter in weight than diesel and are therefore used for small portable mechanical plant such as 50 mm suction pumps, clamp-on concrete vibrators, etc. Diesel is available in two categories:

1. Diesel fuel on which full duty has not been paid; it is marked Gas Oil, coloured red, that is, it contains a dye, and is described by Customs and Excise as 'Rebated Oil'.

2. DERV (abbreviated from diesel engine road vehicle). This is diesel fuel on which duty has been paid and is described by Customs and Excise as 'Unrebated Oil'.

These oils may be supplied in steel drums of 204 litres capacity and currently there is usually a charge of £8.00 on the drum which is recoverable when the empties are returned.

It is intended that rebated oil shall be used only for site-based engines but the law recognises a practical problem where a vehicle such as a site dumper is required to be moved from land in the occupation of the contractor to other nearby land also in his occupation. Provided that the contractor obtains the approval of the Licensing Authority, he may use the dumper on the road with a maximum travel of six miles in any calendar week.

Petrol

The Petroleum Spirits Regulations lay down the conditions for the storage and conveyance of petrol. In every district there is a Petroleum Spirits Officer whose duty it is to enforce the regulation. In some smaller districts the person appointed will combine this with other work such as Weights and Measures Officer. Basic considerations regarding the storage of petrol are:

1. Amount that can be stored in one place.

2. The distance from the road or other premises to the stores.

3. The distance between separate stores of petroleum spirits.

4. The storage arrangements.

5. Fire fighting equipment.

Where a contractor finds it impossible, because of the nature of the site, to comply with the precise conditions in the regulations, the local petroleum spirits officer has the power to waive the regulations and impose new conditions which seem to him appropriate to that site and comply with the letter of the regulations.

The contractor can then opt to either comply with the Petroleum Spirits Regulations or with the conditions imposed by the officer.

Figure 9.4   **Loads carried by vehicles**

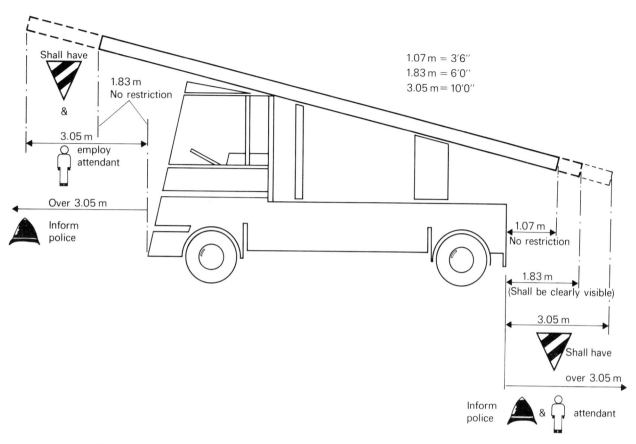

Overall length of loads:
Load on one vehicle — overall length of the vehicle itself plus length of any projection to front and/or rear of vehicle — not to exceed. . . . 18.3 m.

Restrictions on load projections:
To front and/or rear: General rule is that no forward projection shall exceed 1.83 m or a rear projection exceed 1.07 m unless the appropriate requirements of Schedule 8 are complied with as follows:-

	Attendant	Signs	Police
Projection to front:			
Exceeds 1.83 m but not 3.05 m	√	√	√
Exceeds 3.05 m	√	√	√
Projection to rear:		Made visible	
Exceeds 1.07 m but not 1.83 m		√	
Exceeds 1.83 m but not 3.05 m	√	√	√
Exceeds 3.05 m	Specially required	Of specified pattern to be exhibited	Notice to be given

Police: Two clear days notice to Chief of Police of each area in which to be used (excluding Sundays, Bank Holidays, Christmas Day, Good Friday) giving particulars of time, date and route and particulars of load.

Attendants: At least one person in addition to those employed in driving the vehicle to attend the load and give warning to the driver and any other person of likely danger.

Signs: Prescribed signs per Part II, Schedule 8, to be exhibited on relevant projections — one to the front (or rear) and one on each side — kept clear and unobscured. Triangular shape alternate red and white stripes.

Item 12

Part of these regulations deal with projecting loads from lorries. (See Figure 9.4 for details.)

People

**Contracts of
employment**

The regular and punctual attendance of tower crane and excavator drivers is vital to the success of any contract. It follows that the conditions of employment which they are offered should not only secure their attendance but provide motivation for them to get on with the job. Apart from the obvious aspects of security of employment, good pay, bonus, holidays, etc, there is a need to consider a career structure both for mechanics and operators.

Mobile fitters

Large plant hire firms and large contracting organisations usually employ mobile fitters, sometimes equipped with radio control, travelling directly from one site to another carrying out maintenance and reapir. It is obvious that there has to be close liaison between plant yard and site in order to know when the site plant will not be in use and available to the fitters.

Plant engineers

Plant engineers who have high qualities of imagination and innovation can play an important part in winning a contract. Where a construction design is difficult or unusual, the development of a new piece of plant to do the job or the modification of an existing standard piece of plant by the engineers can often win the award of the tender for the company. For such developments a team composed of a site manager, plant engineer, work study officer and estimator collectively provides a lot of expertise in developing new ways of doing work.

Finally, the successful use of plant in the future will probably require materials handling methods to be planned from the merchant's warehouse or factory right through to installation on site.

10. Finance

It is unnecessary to stress the importance of the control of finance; all tangible assets derive from finance. The control of stocks of materials has a direct bearing on finance and so on. The point is obvious. The proprietors of many smaller building firms are often hard working, technically oriented men. In the past, many such firms have prospered and grown large with the proprietor gaining his experience of finance on the way upwards. Today there is a higher element of risk in this practice and the study of the control of finance ranks equally with technical study for the potential builder.

All businesses must keep proper books of accounts. This is mandatory for tax purposes and is, of course, good common sense for the purpose of controlling any business. The principles of financial control are common to most kinds of business, the arrangement of account recording varying with different kinds of business activity. Whilst the ultimate objective of most business activity is to make profit, different kinds of work have different profit potential. The profit potential of a particular type of work may change with a changing work environment. It is foolish to commence a project under-capitalised, no matter how attractive the profit margins appear to be. It is possible for a profitable business to be forced into liquidation through lack of cash. A business with a shrinking work load may have an abundance of cash but be in the midst of a crisis. To be able to judge the correct level of cash liquidity and the amount committed is not easy, and in periods of growth or cut back, it is extremely difficult.

Accounts

Records of all transactions are maintained in accounts. A transaction has taken place when value changes hands. Some transactions are external, for example, when the plant department send a scaffolder to a contract to alter the scaffolding. The plant department has given value, the contract received value. The giver (plant department) must be credited, the receiver (the contract) must be debited. Unless separate accounts are maintained for each external organisation and each internal section, true financial control becomes difficult to apply.

Expenditure

Some expenditure can be directly related to the needs of one particular contract, for example, the payment of wages and salaries to the personnel employed on a contract is clearly related to only that contract. These are 'direct' expenses to that contract. The salaries of the accounts department staff at head office are not directly related to any one single contract and head office expenses such as these can be said to be indirect expenses. The sum of all indirect expenses must be borne by all contracts. Thus, in the price build-up for an estimate, a percentage is apportioned to cover the cost of overhead expenses. Unless proper accounts are maintained of all expenses, it would be impossible to know at what level to set the overheads percentage when preparing estimates for future work.

The classification of ledger accounts

Figure 10.1 illustrates the different sections of the ledger.

Personal accounts

As the name suggests, these are the accounts of persons or firms transacting with the business. This includes the suppliers of materials, sub-contractors, and clients.

Figure 10.1

Diagrammatic representation of the accounts

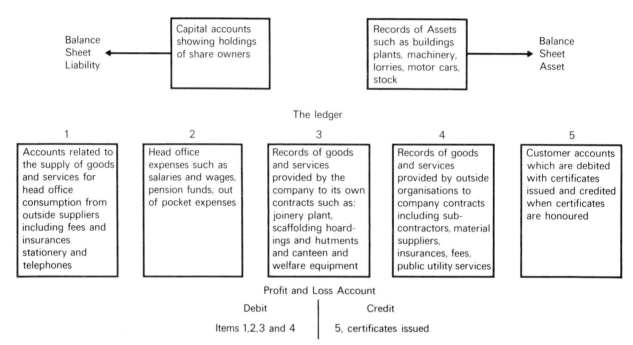

Impersonal accounts

This section can be sub-divided into:

1. Real.
2. Nominal.

Real accounts are those which relate to real things such as buildings, plant, motor cars, office furniture and fittings.

Nominal accounts. Everyday examples of these accounts are: salaries, rent, rates, lighting, heating, discounts, etc.

The double-entry system of book-keeping

Every transaction is recorded twice, once to credit the 'giver' account, the second entry 'debits' the 'receiver' account.

Day books

In order to minimise the number of entries which go into each ledger account, use can be made of day books and in the case of a contract account, the total amount of materials received from outside suppliers in one month may be a single entry in the account, backed up by appropriate pages of computerised details.

Trial balance

At the end of a trading period, the accounts are closed. That is, a balance figure is inserted in each account to create two equal totals. Table 10.1 illustrates this.

Table 10.1

Contract number 7

1980
April 30	£		£
To supplies from various merchants	35 000	By certificate No 4	80 000
To services received from various sub-contractors	20 000	By materials transferred to Contract No 18	2 500
To plant hire charges	10 000	By small tools returned to yard	500
To salaries and wages of directly employed men	5 000		
To balance	13 000		
	83 000		83 000

When all the accounts have been closed, the balance figure is extracted from each account and if all the accounts have been properly kept, the sum of the debit balances should equal the sum of the credit balances.

This act of satisfying that the two totals of balances are equal in amount is known as a 'trial balance'. If the two totals are not equal in amount, then a mistake has been made in entering transactions in the ledger accounts and the bookkeeper must go back and search out the mistake. If the two totals are equal, then the transactions may have been entered correctly in the accounts. However, it should be noted that the trial balance will not reveal errors of omission, ie when the bookkeeper fails to enter both the debit and credit result of a transaction. Nor will it reveal compensating errors, ie when two mistakes occur of similar amount on opposite sides of the accounts.

An example of a simple trial balance

If all figures from 1 to 10 inclusive are placed in two columns in random order, no matter where it is decided to separate the columns into different accounts, the sum of debit balances will equal the sum of the credit balances.

Example 10.1

	10			1
	9			2
	8			3
			Balance	21
Totals	27			27
	7			4
	6			5
			Balance	4
Totals	13			13
	5			6
	4			7
Balance	12			8
Totals	21			21

199

		3			9
		2			10
		1			
Balance		13			
		19			19

Extraction of balance figures

Debits		Credits
12		21
13		4
25		25

The final accounts.

At regular time intervals (which may be three, six, nine, but certainly no greater than twelve month intervals) a trial balance is prepared as a prelude to preparing the final accounts to determine the extent of profit or loss for the period.

The final accounts comprise:

1. Trading account.

2. Profit and loss account.

3. Profit and loss appropriation account.

4. The balance sheet.

The trading account
The trading account is the summation of all the direct expenses of manufacture and all the direct gains from the value of work done. A manufacturer making one product would prepare one trading account. The builder because he may be 'manufacturing' many buildings consecutively, prepares a separate account for each contract. The builder wants to know what profit or loss is being made by each separate contract.

The profit and loss account
This account contains all the profits and losses made on contracts and all the expenses from the 'nominal accounts'. The only gains occurring in the 'nominal' ledger are those in the nature of discounts, etc.

The profit and loss account is the gathering together of all expenses (losses) and all gains (profit). Certain adjustments are shown in the profit and loss account and again in the balance sheet in order that the double entry system is complied with. Typical adjustments relate to the marking down in the value of assets to record as near as possible the true value of the asset to the builder. These might include plant depreciation, office furniture, and an allowance for bad debts. It should be noted that it is not the value of the asset but only the loss in value that is recorded in the profit and loss account.

Profit and loss appropriation account
The ownership of the company may be vested in a number of categories of shareholders. The appropriation account is where the record of the payments made to these diverse groups is noted and also monies placed in 'reserve'. This account is therefore the profit distribution centre.

The balance sheet
This is not an account but is a statement showing the value of the assets and liabilities of the firm at a certain date and time.

The use of the final accounts
One of the main values of the final accounts is to provide a basis for comparing one trading period with another. It will be realised that it is important that the accounts for each period are prepared in exactly the same manner, consistency is very important.

200

Table 10.2   **Simplified final accounts**

Contract Accounts

1. Direct expenses incurred by the firm on behalf of the contract	_____	1. Direct gains by way of certificate values earned	_____
	_____		_____

Profit and Loss Accounts

1. Losses from contracts		1. Profits from contracts	
2. All expenses of running the head office			
3. The marking down in the value of assets	_____		_____
	_____		_____

The Balance Sheet

What the firm owes:		*What the firm owns:*	
1. Capital 'owed' to the proprietors		1. Buildings	
2. Long-term loan from the bank		2. Plant	
		3. Stocks of materials	
3. Materials and services received		4. Work in progress uncertified	
		5. Debtors	
4. Trade creditors	_____	6. Cash at bank	_____
	_____		_____

Table 10.2 shows the general arrangement of the final accounts in simple form. The balance sheet shows the existence of the firm as a separate entity from its owners, the firm 'owes' the capital to the owners.

Example 10.2 extends the treatment shown in Table 10.2 by introducing figures to illustrate the adjustments to the value of assets.

Depreciation of assets.

It is not within the scope of this book to deal with this subject in the depth of treatment which the accountant must give the topic, but the following points should be considered. Three values can be identified:

1. Marking the value down as a percentage of cost price.

2. Deducting an amount deemed to be equal to one year's value at current prices.

3. Making an allowance which is related to replacement cost.

When dealing with plant, it is not always easy to establish what the replacement cost of an asset with three or four years useful life might be because of the rapid rate of technological improvement.

The amount of debtors is £6,000 but the accountant knows from previous experience that approximately 2% of these debts will not be paid. It is unwise to ignore this and the wise man takes a conservative view when calculating profit.

The allowance shown for taxation should be regarded as notional. Tax law is extremely complicated and the amount of tax due on a given amount of profit can vary around a number of factors, not the least of these being the class of business ownership, ie sole trader, partnership or limited liability company.

Contract 7

To supplies from various merchants	35 000	By certificates 1 to 4 inclusive	80 000
To services received from various sub-contractors	20 000	By materials transferred to Contract 18	2 500
To salaries and wages of directly employed	5 000	By small tools returned to plant yard	500
To plant hire charges	10 000		
To balance to profit and loss a/c	13 000		
	83 000		83 000

Contract 11

To supplies	10 000	By penultimate and final certificate	14 000
To sub-contractor	2 550	By profit and loss a/c	1 350
To salaries and wages of directly employed	2 400		
To plant hire charges	250		
To local authority and statutory body fees	150		
	15 350		15 350

Profit and Loss A/c

To plant depreciation	300	By Contract 7	13 000
To Contract 11	1 350		
To offices salaries	6 000		
To lighting and heating	1 000		
To hire of staff cars	1 000		
Gross profit to Profit and Loss Appropriation	3 350		
	13 000		13 000

Profit and Loss Appropriation A/c

To provision of bad debts	150	By profit and loss a/c	3 350
To provision of taxation	1 000		
To reserve	1 000		
To net profit	1 200		
	3 350		3 350

202

Owner's Capital A/c

To balance c/d	14 550	By balance b/d		13 350
		By profit appropriation		1 200
	14 550			14 550

The Balance Sheet   Date . . . .

Capital	13 350		Buildings		10 000
Add profit	1 200	14 550	Plant	3 000	
Reserve fund		1 000	*less* depreciation	300	2 700
Bank overdraft		2 000	Debtors	6 000	
Provision for tax		1 000	*less* provision	150	5 850
Creditors		4 000	Cash at bank		4 000
		22 550			22 550

Business ownership.

Sole trader

A sole trader 'enjoys' unlimited liability, that is, in the event of business failure, his creditors may expect to claim against his personal property. Such a person takes all the profit.

Private company

A sole trader who converts his business into a private company relieves himself of the burden of unlimited liability, but in so doing he becomes a less attractive debtor to a bank or other normal source of business finance. In order to have his former borrowing powers, he may find it necessary to offer his house as security for loans.

Partnerships

Partnerships in trade organisations are now much less common than formerly. It should be noted that the apportionment of profit or loss can be in whatever proportions the partners decide. Like the sole trader, the partners involved in the day to day running of the firm have unlimited liability for the firm's debts.

Private limited company.

A private company limited by shares may be formed by two or more persons with an upper limit of 50. Once in existence, the upper limit should not exceed 50 excepting for former employees who may remain as shareholders. Limited companies must be registered with the Registrar of Companies. A registered company is a legal entity and despite death and retirement of founder members, there is no legal reason why it should not continue after all the founder members have ceased to exist.

Public company

A successful private company may find after a time that its growth and development is restricted through lack of capital. In this event, the owners may decide to make an application, by filing a statutory declaration, to convert the organisation into a public limited liability company. The public at large may see a notice in the press inviting applications to buy share capital. The would-be investor will be advised to send a sum in respect of each share he wishes to purchase. This may be in the order of 20p in respect of each £1.00 ordinary share. This is known as allocation money. Should the share issue be oversubscribed, it will normally be rationalised on a first come first served basis, or on a proportionate issue basis. If the issue is undersubscribed, the residue will be purchased by the underwriters who would sell them at an appropriate future time.

Calls

An investor, having received his share certificate and had his name entered on the Register of Shareholders, will later be asked to pay a further sum in respect of each share held. This is known as the first call. Later, the investor may be asked to pay second and third calls, but the most he will pay in tote will be the face value of the share. This is the extent of his liability. Some shares are never fully paid up. The holder of such a share is liable to be asked for the outstanding amount at any time. Payment by stages in this way enables the new company gradually to get the money working.

Once a company is successfully floated, it always has shareholders. In order to sell a share, the holder must find a buyer.

Categories of shares.

Ordinary shares

The holders of these shares are the true owners of the company and are generally the only group of shareholders entitled to vote at shareholders' meetings. Ordinary share capital is referred to as the equity capital. The dividend paid to ordinary shareholders varies according to the prosperity of the company.

Preference shares

These shares bear a fixed rate of interest and they are named preference because they have preferential treatment in the payment of their interest over the ordinary shareholders' dividend payment and also in repayment of capital on the liquidation of the company.

Cumulative preference shares

These shares have a priorty over preference shares and in the event of no profit being made to pay the fixed rate of interest, the cumulative preference shareholders get the backlog when future profits permit. The fixed rate of interest is lower than that of preference shares because of the greater certainty offered.

Loans

Debenture Stock

These are fixed interest loans. As such, the interest is an expense of running the business and has a prior claim over dividends to shareholders. These stockholders have no voting rights unless the company default in the payment of interest, in which circumstances they would usually acquire that right. Debenture stock may be redeemable at par (face value) in a stated period of years or they may be irredeemable. If they are to be redeemed, the Board must ensure that adequate funds are available at the time of redemption.

Statutory auditors.

It is mandatory for every limited company to appoint external professional auditors. Their duties are defined in the Companies Acts of 1948 and 1967. The main reason for their existance is to protect the interests of the shareholders. Apart from direction given in the Acts about the Report (S14 1967), an auditor is expected to approach his work in the spirit and manner laid down by the professional institutes. If the final accounts and balance sheet are not properly drawn up, he must report the matter to the shareholders. If he does not do so, he is guilty of wrongful exercise of lawful duty. Above all things, an auditor must be honest, he must not certify what he does not believe to be true.

However, he is not guilty of negligence if he accepts the word (in writing) of a responsible official that the value, say, of the timber stock is 'x' thousand pounds. He is not a stock-taker and providing there is nothing to arouse his suspicions, he cannot be personally responsible for matters such as this. However, in the United States, auditors do test check actual physical stock. It is therefore possible that United Kingdom auditors may do this in the future.

Capital and Liquidity.

Liquidity.

Money in the bank is liquid, money tied up in buildings or other similar fixed assets is far from liquid; hence the name 'fixed' asset.

Fixed assets are necessary to provide a base from which the company operates. Current assets support the level at which operations are taking place.

Raising cash.

A company may consider raising cash for a number of reasons. Some suggestions are as follows:

1. It may have under consideration a capital expenditure scheme.

2. It may be able to forecast a cash famine some months ahead due to a coincidence of contract starts.

3. It may wish to expand.

4. The directors may have been imprudent and allowed a drift to occur of current assets to fixed assets.

5. A cash shortage may be foreseen due to the company having a significant value of contracts which have slower arrangements for paying for work than normal.

Raising cash internally

1. If a firm saves up to pay its interim dividend of around 5% in October and a further 15% or so in April, this saved money could be used, for a brief period only, to provide short-term finance.

2. In a somewhat similar way, a firm may save by way of depreciation charges to replace plant and other assets, but many contractors now hire in much of the plant and equipment which they use.

3. When the demand for contractors' services is strong, the negotiators should try to arrange with clients conditions of payment which are speedier than normal, or with much the same effect, negotiate better conditions of credit from suppliers.

With reference to Table 10.3, it can be seen that almost 50% of the assets are tied up in freehold buildings. When considering ways by which the firm may be able to improve its liquidity through its ownership of the freehold building, the value of the building shown in the balance sheet should be examined. It may be open to three interpretations:

1. It may be related to the original purchase price less annual deductions for depreciation.

2. Because of inflation, the value in the books has been regularly updated to reflect the current market value.

3. The book value is based on an assessment of the service the firm obtain from the use of the building.

Raising cash externally

The company may be able to move to another building, the rent or lease of which is less than the rent or lease which they gain from letting or leasing their own premises. In any calculations, repair costs, management expenses, and the cost of the move, must be calculated. It may be quite impractical to move as mentioned and the speed of improvement in the liquidity position by this method may be far too slow. In this event, one of the following methods may be feasible:

1. To make an arrangement with an institutional investor such as an insurance company to enter into a 'sale and lease back' agreement. The lease would normally be for a minimum of twenty years and the annual rental sum in the order of six to ten percent of the agreed sale price. This would be a fixed amount for the period of the lease.

2. The alternative to this would be to raise a mortgage on the premises and then buy back outright in an agreed period; twenty years would be a normal minimum. Mortgage interest rates float with the movement of the minimum lending rate so that the annual outgoings could vary quite considerably. If the mortgage included the industrial part, ie the plant yard, etc, the amount raised would be considerably less than its value as a going concern. The reason for this is because the buildings involved would have limited application to other industries and the lender would wish to protect himself against the possibility of the builder going into liquidation. The net result of this is that the builder is underselling his security.

Investment allowances and Taxation

Before selecting a method of raising money, it is important to look at the central government inducements regarding investment acquisition, etc., which prevail at the time and also at the detailed taxation implications of the various alternatives.

If a public company has a good profit record, it may choose to issue a 'rights issue' of ordinary shares. These are first offered for sale in proportion to their holding, to existing shareholders, for example shareholders are offered one new share for every two existing ones held. If all shares are taken up by existing shareholders in the proportion in which they held shares the voting pattern is unlikely to be disturbed. However, a 'rights issue' in a private company may not be so successful from this latter point of view. It might be that a number of the proprietors might not wish to avail themselves of the offer. In this event, it may be necessary to make an approach to one or two selected outsiders, taking care to assess where the balance of power would lie should they join the firm.

Borrowing from the bank.

Banks are intermediaries, they borrow from the public at large and lend to selected customers. If some of the borrowing customers default, the bank has still to pay the interest on the money to the depositor and ultimately repay his capital. Not surprisingly, they are very selective in choosing borrowing customers. A bank manager is not unduly impressed by the prospective borrower who explains about the very high rate of return which might be earned from the proposed loan, since he is only being asked to share the risk but not the bonuses. Traditionally, banks have been regarded as providers of short-term loans in so far as small firms are concerned. They have tended to move from this position in recent years towards medium and long term loans. For the purposes of borrowing one to two years can be considered short-term, five to seven years medium, and more than ten years long term.

In seeking security for loans, banks would not be very interested in lending against typical builders' stock in the yard which 'turns over' very slowly. Builders' plant may be more interesting, providing the section is well organised commercially. Against buildings used for the conduct of business, they may lend up to about one third of the value. This may seem a small proportion, but it must be borne in mind that this is not self-liquidating finance and of itself provides no evidence that the builder will ever be able to repay the loan. Buildings being erected as a commercial development are a different proposition. There can be considerable security in the value of land, the work in progress, and the stock which is quickly going to be turned over. Assuming the builder owns the land and offers this as security, an arrangement for cash to be borrowed against the 'measure' in the external works could be made. Money for individual houses or dwellings would be available in the form of stage payments, possibly four stages, such as:

1. When work is up to dpc.

2. First floor joist level.

3. Wall plate level.

4. On practical completion.

An agreement would be made about monitoring the progress either through the architect or an independent quantity surveyor.

Most banks would expect the builder to produce ample evidence when negotiating the loan that he had given adequate thought to the planning processes (which is rather different from saying prepared an overall programme). They would expect evidence that the builder had some idea where the customers would come from and that the type of houses offered would have adequate appeal. They are interested in seeing the full order book aspect of the business.

Type of loans from banks

Overdrafts

One of the most common arrangements is to negotiate overdraft facilities up to an agreed maximum. The interest charges are calculated daily on the amount outstanding each day. The interest charges are related to the minimum lending rate, invariably more than this. The amount charged over the rate will largely reflect how the banker feels about lending money to that person or organisation. The typical range would be 1% to 5% above the minimum lending rate. Since the minimum lending rate is used as a regulator by the Bank of England, and since international exchange rates are extremely volatile, the rate tends to change very frequently.

Because the rate changes rapidly, estimates of the profit which might be earned from the use of the money should be conservative. The major drawback is that the bank will reserve the right to withdraw the overdraft at short notice. The bank likes to feel that the overdraft is 'topping up' the firm's finances and does not want it to be regarded as part of the fixed capital structure.

Term loans

If the firm is well established in its mode of operation and soundly based financially, the bank may be interested in providing 'fixed' capital. This would be for a fixed period of years at a fixed interest rate and the security would need to be demanding upon the builder. This might include the careful monitoring of the firm's affairs by the inclusion, for the period of the loan, of a member of the bank staff on the board of directors. Directors' guarantees provided by the chief executive and others, against their personal effects may also be required.

Short term loans.

When a builder wishes to buy something for a definite known amount, such as a fork lift truck or a piece of office equipment, he may approach the banker for a short-term loan of a fixed sum of money. The builder will compare this against the hire purchase scheme offered by the vendor of the machinery and these days, as offered by the banker himself.

General observations concerning banks.

Whilst bankers obviously want to be sure that the would-be borrower has carried out a thorough investigation into his opportunities and needs, there are two other important considerations:

1. The input of the borrower into the project must be a greater amount than the loan requested.
2. The banker lends to the man. That is, he wants to be sure that he is lending to a man who has the type of personality to work hard and of sufficient courage and optimism to see things through to a satisfactory conclusion.

Balance sheet ratios.

A number of ratios can be developed from the balance sheet and the final accounts for comparing the current year's trading with previous years. It is very important that there should be absolute consistency in the preparation of the accounts and the application of the ratios. When carrying out an intra-firm comparison of this kind, it is important to recognise that a change in a ratio is a signal to look deeply before jumping to any conclusions, nothing more.

The balance sheet is a statement of the affairs of a firm on a particular day. This point is illustrated by Tables 10.3, 10.4 and 10.5. In practice shareholders' dividend is usually paid in two stages, an interim payment at the half year stage (typically, this would be around one third of the forecasted dividend for the year), the remaining two thirds or so being paid at the end of the trading period.

Table 10.3

Balance Sheet at 31 December 1985

Issued Capital		Fixed Assets	
50 000 8% Redeemable Cumulative Preferance shares of £5.00 each	250 000	Freehold buildings	500 000
		Plant	100 000
500 000 Ordinary shares of £1.00 each	500 000	Lorries and cars	90 000
Reserve	95 000		
Current Liabilities		Current Assets	
Dividend to shareholders	125 000	Timber stock	24 000
Inland Revenue	15 000	Trade debtors	52 000
Trade Creditors	35 000	Cash	254 000
	1020 000		1020 000

Table 10.4

Balance Sheet at 1st January 1986

Issued Capital		Fixed Assets	
50 000 8% Redeemable Cumulative Preferance shares of £5.00 each	250 000	Freehold buildings	500 000
		Plant	100 000
		Lorries and cars	90 000
500 000 Ordinary shares of £1.00 each	500 000		
Reserve	95 000		
Current Liabilities		Current Assets	
Inland Revenue	15 000	Timber stock	24 000
Trade creditors	35 000	Trade debtors	52 000
		Cash	129 000
	895 000		895 000

Table 10.5

Extract from Profit and Loss Appropriation Account at 1st January 1986

Net profit	400 000
Corporation Tax at 40%	160 000
	240 000
8% Preference share dividend	20 000
	220 000
Ordinary share dividend at 20% gross	125 000
Undistributed profit	£95 000

The acid test ratio $= \dfrac{\text{liquid assets}}{\text{current liabilities}}$

At 31 December 1985

$$\frac{306\ 000}{175\ 000} = 1.7:1$$

At 1 January 1986

$$\frac{181\ 000}{50\ 000} = 3.6:1$$

Analysis.

The position on the 1st January is different from 31st December because shareholders dividend has been paid. This demonstrates the importance of being consistent in the selection of dates for ratio analysis. Although the acid test ratio appears much improved the total position of the firm remains much the same.

Availability of funds $= \dfrac{\text{current assets}}{\text{current liabilities}}$

At 31 December 1985

$$\frac{330\ 000}{175\ 000} = 1.91:1$$

At 1 January 1986

$$\frac{205\ 000}{50\ 000} = 4.1:1$$

Dividend cover.

This measures the number of times that the ordinary share dividend is covered by the earnings of the trading period.

$$\frac{£220\ 000}{£125\ 000} = 1.76:1$$

Dividend yield.

It is assumed that the ordinary shares were being bought and sold at £2.50. In this ratio, the dividend income is measured as a percentage of the real value of the share. Assume a dividend of 20%; this relates to the face value of the share.

$$\frac{£1.00 \text{ x } 20\%}{£2.50} = 8\%.$$

Each share earns 20p. This earning is related to a share value of £2.50 or 8% in the £1.00. For a more detailed explanation of dividend yield see the section Investment Appraisal.

Price/Earnings.

This ratio is related to the share value and the amount earned by the share in the trading period.

It should be noted that after the fixed rate of dividend had been paid to the preference shareholders, there remained £220 000. Once the dividends have been allowed for the cumulative preference and preference shareholders, the remaining profit is deemed to have been earned by the ordinary share capital.

$$\frac{\text{Value of shares at £2.50 each}}{\text{profit earned}} = \frac{£1\ 250\ 000}{£\ \ \ 220\ 000} = 5.67 : 1$$

Operating profit/turnover ratio

$$\frac{\text{operating profit}}{\text{turnover}} \text{ x } \frac{100}{1}$$

The application of this ratio to building work should be treated with caution. The relative proportions of work done by sub-contractors and that by directly employed men is an important consideration. Sub-contracted work usually requires much less funding than that carried out by directly employed workers. Risks are usually less and profit margins lower; thus, if there is a move away from direct employment, total workload needs to be increased to maintain profit level. This ratio is used to measure the adequacy of profit margins, namely operating profit.

Capital

Return on capital employed

Capital employed is deemed to be the sum of assets minus liabilities towards outside parties. It is advisable to consider profit before tax since the rate at which tax is levelled varies and would distort the comparison, thus:

$$\frac{\text{operating profit before tax}}{\text{total assets} - \text{liabilities to outside parties}} \text{ x } \frac{100}{1}$$

$$\frac{400\ 000}{1\ 020\ 000 - 50\ 000} \text{ x } \frac{100}{1} = \frac{400\ 000}{970\ 000} \text{ x } \frac{100}{1} = 41.2$$

Capital Gearing

This is the ratio between the ordinary shareholding and the fixed interest shareholding. The following figures can be derived from Table 10.3.

$$£500\ 000 \ : \ £250\ 000$$
$$2 \ : \ 1$$

When the amount of fixed interest dividend capital is considerable, relative to the amount of ordinary share capital, the company is said to be highly geared. The weakness of high gearing becomes apparent when a company is exposed to bad trading conditions. The following example illustrates this point:

Table 10.6 Gearing

Gear-ing	Ordinary share capital	Fixed interest capital at 8%	Profit for the year	Fixed interest capital share of profits		Ordinary share capital share of profits	
	£	£	£	Amount	%	Amount	%
	500,000	250,000	45,000	20,000	8	25,000	5
Low	500,000	250,000	70,000	20,000	8	50,000	10
	500,000	250,000	95,000	20,000	8	75,000	15
	500,000	250,000	32,500	20,000	8	12,500	2½
	250,000	500,000	45,000	40,000	8	5,000	2
High	250,000	500,000	70,000	40,000	8	30,000	12
	250,000	500,000	95,000	40,000	8	55,000	22
	250,000	500,000	40,000	40,000	8	—	—

If the return on capital employed is calculated for a company with a total capital of £750 000 and earning a profit of £45 000

$$\text{The return on capital} = \frac{45\ 000}{750\ 000}$$

$$= 6\%.$$

However, in the last line of Table 10.6, it can be seen that when the profit falls to £40 000 (approx 5½%), with capital gearing as shown and an 8% fixed dividend, the profit for ordinary shareholders is zero. However, with a relatively slight improvement in the performance of a highly geared company potential dividends to ordinary shareholders rise sharply. £70 000 profit is 8% on total capital but provides a return of 12% for ordinary shareholders.

Turnover of capital

A company with a capital of £10 000 and invoicing work of £100 000 in one year is said to have turned its capital over 10 times.

Interfirm Comparison

All the analysis considered so far has been about one firm comparing its own present performance with its past performance. If the firm in question happens to be a very poor performer, it is doubtful whether it is of value to know that the results are slightly better or worse than awful! There seems to be a case for both interfirm comparison and intra firm comparison.

The Centre for Inter-firm Comparison

This is an organisation set up in 1959 by the British Institute of Management in association with the British Productivity Council to meet the demands of trade and industry for an expert body to conduct interfirm comparisons on a confidential basis as a service to management.

The Centre was established as a separate company in order that its services should not be confined to members of its sponsoring organisations but made widely available to industry and trade. It depends for its income on fees charged for its services, but it is a non-profit-making organisation, and its charges are therefore based on actual costs incurred. Its executive staff has wide experience of interfirm comparison in more than 60 industries, and are therefore well equipped to deal with any problems which may arise in an industry where comparisons have not previously been conducted.

How an interfirm comparison works

Several firms of an industry or trade contribute their figures confidtially to the Centre, from which they receive in return a report which shows, anonymously and in the form of ratios or percentages, the data of each participating firm. The report also explains how the results are to be used by the individual firm. In most IFC schemes each firm taking part receives a confidential individual report. This discusses its position, high-lights points of weakness or strength shown up by the IFC, interprets differences between the firm's ratios and those of others, and indicates the directions in which improvements should be made. In many IFC schemes arrangements are made for the Centre's staff to discuss this report with the company's management.

As the above description and Table 10.5 show, an IFC is not a statistical survey, but an important aid to self-diagnosis for each of the firms taking part.

Interim comparison and the building industry.

Whether the building industry will become actively involved with interfirm comparison in the future remains to be seen. So far, the response has been lukewarm.

INVESTMENT APPRAISAL

Yield

Individuals or corporate bodies may wish to buy shares in a company for any one or a combination of the following reasons:

1. To own sufficient shares in the company in order to control its direction and policy.

2. To have an income from the shares in the form of dividend.

3. To invest in shares which are likely to increase in value and thus sell them at some time in the future for a profit.

Before dismissing point number one because it is not relevant to the topic, it can be mentioned that if a person owns 51% of the equity shares in a company, he clearly controls the company. In practice, one body owning as little as 25% might control providing the remainder of the shares were in diverse hands in small lots.

With regard to points two and three they cannot be separated because they both concern the same thing, namely, return on investment. At any one time there are three factors relevant to investment value:

1. Par value (face value)

2. Current value

3. Current rate of dividend.

From these three factors a basis of comparative earning power between shares can be developed; this is known as the yield.

1. Assume some £1.00 ordinary shares are currently earning 18% dividend and selling in the market at £1.50. Yield can be calculated as follows:

Par value x dividend = Market price x yield

£1.00 x 18 = £1.50 x yield

$$\text{Yield} = \frac{1.00 \times 18}{1.50}$$

$$= 12\%.$$

2. If a 25p share is selling in the market for 60p and earning a dividend of 20%, what is the yield?

$$25 \times 20 \quad = 60 \times \text{yield}$$

$$\frac{500}{60} \quad = \text{yield}$$

$$8.33\% \quad = \text{yield}.$$

3. Assume a £1.00 share is selling in the market for 80p and earning a dividend of 9%, calculate the yield:

$$£1.00 \times 9 \quad = 80 \times \text{yield}$$

$$\frac{900}{80} \quad = \text{yield}$$

$$11.25\% \quad = \text{yield}.$$

In order to appreciate the significance of yield, the reader should ask the question, if he/she had £12.00 to invest, which of the following shares would be purchased?

Type 1. These £1.00 ordinary shares are selling at £1.50, so 8 shares could be purchased. They are currently earning 18% of face value, so 8 x 18p = £1.44.

Check: $\dfrac{£1.44}{£12.00} \times \dfrac{100}{1} = 12\%$ yield.

Type 2. These 25p shares are selling at 60p, so 20 shares could be purchased for £12.00. They are currently earning 20% of face value which is 5p each: 20 x 5p = £1.00.

Check: $\dfrac{£1.00}{£12.00} \times \dfrac{100}{1} = 8.33\%$ yield.

Type 3. These £1.00 shares are selling at 80p, so 15 shares could be purchased for £12.00. They are earning a dividend of 9% of face value.

Dividend = 15 x 9p = £1.35.

Check: $\dfrac{£1.35}{£12.00} \times \dfrac{100}{1} = 11.25\%$ yield.

Rank order of preference is 1,3,2.

Discounted cash flow.

This provides a useful basis for the comparison of projects. The underlying principle is that the value of all outflows and inflows of money from the projects are brought to a basis of a value *now*. Money is worth more now that an equal and certain sum in the future because it can be invested in the period between. As a starting point, if the present value of all inflows exceeds the present value of all the outflows, both inflows and outflows having been discounted at the forecasted average interest for the period of the project, then the project is profitable.

Example 8.3   For the purpose of this example, a notional interest of 10% per annum is assumed.

The difference between these two projects is insignificant. B might appear, at first glance, to be much better because the cash inflows are £2 000 more.

Project A — **Initial expenditure, time now £15 000**

Cash inflows (£)	Time	Equal to	Value now (£)
5,000	End of year 1	$5,000 \times \dfrac{1}{1.1}$	4,545
5,000	,, ,, ,, 3	$5,000 \dfrac{1}{(1.1)^3}$	3,757
5,000	,, ,, ,, 5	$5,000 \dfrac{1}{(1.1)^5}$	3,105
5,000	,, ,, ,, 7	$5,000 \dfrac{1}{(1.1)^7}$	2,566
5,000	,, ,, ,, 9	$5,000 \dfrac{1}{(1.1)^9}$	2,120
$\dfrac{5,000}{30,000}$	,, ,, ,, 10	$5,000 \dfrac{1}{(1.1)^{10}}$	$\dfrac{1,928}{18,021}$

Value of investment	15,000
Present value of return	18,021
Return on investment based on present values	3,021

Project B — **Initial expenditure, time now £15 000**

Cash inflows (£)	Time	Equal to	Value now (£)
1,000	End of year 1	$1,000 \times \dfrac{1}{1.1}$	909
5,000	,, ,, ,, 3	$5,000 \times \dfrac{1}{(1.1)^3}$	3,757
9,000	,, ,, ,, 5	$9,000 \times \dfrac{1}{(1.1)^5}$	5,588
9,000	,, ,, ,, 7	$9,000 \times \dfrac{1}{(1.1)^7}$	4,618
2,000	,, ,, ,, 9	$2,000 \times \dfrac{1}{(1.1)^9}$	848
$\dfrac{6,000}{32,000}$	,, ,, ,, 10	$6,000 \times \dfrac{1}{(1.1)^{10}}$	$\dfrac{2,314}{18,034}$

Value of investment	15,000
Present value of return	18,034
Return on investment based on present value	3,034

Perhaps the greatest value of comparisons such as these is to highlight the factors which have not been considered.

1. Inflation rate affecting both projected cost and inflow of funds.

2. No attempt has been made to quantify the risk associated with each project.

3. Associated with item 2, the type of management expertise required for each project and if the requirements differ, which is the more desirable from the viewpoint of using up readily available resources.

4. It would be extremely difficult to make an assessment of the level of tax payable for each of the next ten years and although each project is planned for ten years, the payments to the Exchequer might vary considerably for the two projects because income patterns are very different.

Discounted Cash Flow is a useful tool but it has limitations which must not be ignored.

Cash Flow Forecasting.

All businesses must operate some form of cash flow forecasting. Cash flow forecasting is probably as difficult for a housing developer as any other form of business. From buying the land to selling it in small packages to individual house purchasers, a considerable time can elapse. The cost of land can amount to 40% of the total cost and it is the first purchase that the speculator must make. Pieces of land available for housing development vary in price about a number of factors:

1. The desirability of the property from residential aspects of the location.

2. Proximity to roads, railway and bus services, schools, etc.

3. Size. Large pieces of land attract fewer buyers, not all firms can afford to 'tie-up' large sums of money, the price per acre may tend to be less for large plots of land.

4. Topography. A level site is easier to develop than a sloping site.

5. Obstructions. A site which is easy of access, containing no old buildings or large trees, is cheaper to develop.

6. Availability of services. Water, gas electricity, telephones.

7. Demand and availability.

The developer is always busy looking for land. He must have enough land purchased to keep the production flowing at the required rate of 'x' houses per annum. When he receives particulars of two or three plots of land, he must consider the type of houses suitable for each location, planning densities and finally input in cash and output. The input and output of cash will be related to his overall programme of construction; and from this he will extract a cost/income programme.

Table 10.7 **Cost income programme for a private development of 20 houses**

Item	O	N	D	J	F	M	A	M	J	Jy	A	S	O	N	D	J	F
Purchase of land	X																
Solicitor's fees			X														
Roads and sewers				X	X	X	X										
Foreman ganger, unloading						X	X	X	X	X	X	X	X	X	X		
Materials and labour						X	X	X	X	X	X						
Labour and materials s/c									X	X	X	X					
Garages and other external works												X	X				
Estate agent and solicitors fees												6	6	8			
Ancillary costs					X							X					
Head office overheads					X			X				X					
Sales												6	6	8			
Maintenance																X	
Total charges or gains (40)	1	0	1		2	3	3	3	3	3	3	3	7	4	3	0	1	0

From his prior knowledge, enquiries and studies, figures as shown in Table 10.8 can be produced.

Table 10.8

Cost breakdown for one house and 20 houses on a private development of 20 houses

	1 house (£)	20 houses (£)
Materials	2 375	47 500
Labour	1 500	30 000
Labour and materials s/c	2 000	40 000
Garages	400	8 000
	6 275	
Preliminaries:		
Foreman, ganger, unloading	1 250	25 000
Roads and sewers	875	17 500
Other external work	325	6 500
	8 725	
Land	6 810	136 200
Estate agent	375	7 500
Solicitors	120	2 400
Ancillary costs	75	1 500
	16 105	
Head office overheads 5% of construction costs	435	8 700
	16 540	
Interest on capital at 10%	___	
Maintenance	100	2 000
Profit at 10% on all costs	___	
Sale price	___	

NB Interest on borrowed capital is calculated at a notional figure of 10%.

Using these two documents, the projected cash flow statement (Table 10.9) is produced.

Explanation of projected cash flow

The solicitors' costs are deemed to occur when the land is purchased in October (£1,200) and at the rate of £60 per house when the houses are sold. The allowance for interest on capital is calculated as follows:-

1. Calculate the sum of the debit balances.

2. Find the average debit balance.

3. Multiply the average debit balance by 10% and the period in years of indebtedness.

1.	October	136 200
2.	November	136 200
3.	December	137 400
4.	January	144 675
5.	February	152 300
6.	March	172 095
7.	April	191 890
8.	May	210 210
9.	June	235 630

			10.	July	262 050
			11.	August	286 470
			12.	September	204 170
			13.	October	102 895

$$£2372\ 185 \div 13 = £182\ 476$$

$$\text{Interest calculation} = \frac{£182\ 476}{1} \times \frac{13\ \text{months}}{12} \times \frac{10}{100}$$

$$= £19\ 768$$

Table 10.9   Projected cash flow for a private development of 20 houses

Month	Item	Amount	Debit Month	Cumulative	Credit Month	Cumulative	Bank —	+
Oct.	Land	136200	136200	136200			136200	
Nov				136200			136200	
Dec	Solicitors	1200	1200	137400			137400	
Jan	Roads & sewers	4375						
	Head office overheads	2900	7275	144675			144675	
Feb	Roads and sewers	4375						
	Foreman, garages, unloading	2500						
	Ancillary cost	750	7625	152300			152300	
Mar	Roads and sewers	4375						
	Foreman, garages, unload	2500						
	Materials and Labour	12920	19795	172095			172095	
April	Roads and sewers	4375						
	Foreman, garages, unload	2500						
	Materials and Labour	12920	19795	191890			191890	
May	Foreman, garages, unload	2500						
	Materials and labour	12920						
	Head office overheads	2900	18320	210210			210210	
June	Foreman, garages, unload	2500						
	Materials and Labour	12920						
	Labour and materials s/c	10000	25420	235630			235630	
July	Foreman, garages, unload	2500						
	Materials and Labour	12920						
	Labour and materials s/c	10000	25420	262050			262050	
Aug	Foreman, garages, unload	2500						
	Materials and Labour	12920						
	Labour and materials s/c	10000	25420	286470			286470	
Sept	Foreman, garages, unload	2500						
	Labour and materials s/c	10000						
	Garages & other ext. work	7250						
	Estate agent & solicitor	2900						
	Ancillary costs	750						
	Head office overheads	2900	26300	312770				
	Sales				108000	108000	204170	
Oct	Foreman, garages, unload	2500						
	Garages & other ext. work	725						
	Estate agent & solicitor	2900	6125	318895				
	Sales				108000	216000	102895	
Nov	Foreman, garages, unload	2500						
	Estate agent & Solicitor	2900	2790	321685				
	Sales				144000	360000		38315
Dec	Nil							
Jan	Maintenance	2000	2000	323685				
	Interest charges		19768	343453				16547

217

It should be noted that the profit of £28 518 is earned in 13 months. The profit rate per annum is £26 319. The percentage profit on capital is:

$$£ \frac{15\,276}{182\,476} \times \frac{100}{1} = 8.3\%.$$

Reliability of planned cost figures

It is always difficult to predict the cost of external works. Uncertainty of what obstacles might be met with in the ground makes accurate predictions difficult. Once the work is brought up to ground level, site managers involved in repetitive house building can predict labour and machine hours input with a high degree of accuracy. The surveyor's chances of forecasting labour and material cost is not quite so high. Another main area of uncertainty is the forecasted sales programme. People buying a house may have a hold up in selling the one they occupy, sales falling through just before the customer is due to 'sign on the line' happen all too often. Also, labour rates on housing work can fluctuate rapidly. Because productivity varies very much with the weather, many developers try to get the roofs on houses under construction before Christmas, avoiding starting any new construction until March. This tends to make an unequal demand pattern for the labour. The developer who chances the weather, ie increases the risk of delay, defective work, etc, may buy labour marginally cheaper than those who 'play safe'.

FINANCIAL CONTROL OF CONTRACTS

There are many different types of contractual arrangement possible between client and contractor. Both client and contractor have to control the financial aspects of a project. This section looks at the problem from the contractor's side and assumes the use of the JCT Form of Contract unless stated otherwise.

The starting point of financial control is to prepare a production programme. The programme shows what work will be done, and when, and the resources of men, machines, and materials required to achieve it.

Wages

With regard to directly employed men and labour only sub-contractors, it is possible to predict with a good deal of precision what the cash outflows will be, providing the production programme is adhered to. Wages paid to directly employed men in the first week of each certification period will usually be recovered from the employer after five or six weeks have elapsed. As certificates are paid at monthly intervals, the adverse effect of the six week delay is felt in the first period and is an ongoing two week forward input by the contractor.

Plant

Plant owned by the main contractor should be properly invoiced by the plant department to the contractor at monthly intervals which coincide with the certificate periods. Plant owned by outside hirers will be invoiced monthly and paid by the accounts department.

Materials

Materials are generally supplied on credit, the time period allowed for the credit and any cash discounts available will depend upon the financial stability of the contractor, the value of the order and the desirability of future orders, current trading conditions and finally, the negotiating skill of the contractor's buyer. Materials supplied by merchants will generally be paid for within two months of being received on site. If the contractor keeps a close control on stock levels, it is possible that he will not make any significant input of cash to cover materials of construction.

Where components, such as concrete cladding panels, are manufactured in offsite workshops, they may be included in certificates provided they are labelled or suitably catalogued and patently intended for delivery to that particular site. Overordering and overstocking leads to financial ineffic-iency. Materials ordered over and above actual requirements are seldom converted back into full cash value. Overstocking means that there is a stock of cash equivalent to the value of the excess stock stood idle.

Responsibility for financial control at site level.

Site managers have primary responsibility to ensure that there are no idle resources on site. If they have men or plant idle, they should either organise their removal from site or put them to work. Forward planning is a necessary preliminary to a successful financial outcome.

The contract surveyor should keep on top of financial administration so that all items which can be legitimately included in certificates are included. The speedy pricing of variations is important and when this work does get behind, pricing out of logical sequence to get those converted to cash which give the biggest return.

Cost/value comparison

In any cost/value comparison due account must be taken of any 'weight-ing' which has been applied to the Bill of Quantity rates. 'Front-end weight-ing' gives higher profit margins than normal on selected Bill items which are to be carried out early in the construction programme and consequently lower margins are achieved on some later Bill items. The intention is that the tender should be competitive, win the day, but lead the contract towards a self-financing state earlier in the contract than would otherwise occur.

Table 10.10   **Weighting the bill**

Bill Ref	Details	Quantity	Labour		Plant		Material		Subcontractor		Profit		Completed Rate		+/−	Market Rate
			Value	£	Value	£	Value	£	Value	£	Value	£	Value	£		
E1	Bulk excavation	10000m³	0·50	5000	0·40	4000	0·05	500	—	—	0·10	1000	1·05	10500	+	1·00
B1	200 mm Clinker concrete block walling	1000m²	7·00	7000	0·50	500	5·00	5000	—	—	0·20	200	12·70	12700	+	12·30
P3	7·9 mm Browning undercoat	4000m²	2·00	8000	0·20	800	3·50	14000	—	—	0·05	200	5·75	23000	−	6·0
P3	1·6 mm Finishing coat	4000m²	1·20	4800	0·20	800	2·00	8000	—	—	0·05	200	3·45	13800	−	3·50

N.B. The sum of the positive weightings approximates to the sum of the negative weightings in total value pounds sterling

Table 10.11 shows how the projected cash flows for all contracts can be consolidated onto one statement to show the overall effect of the firm's cash position. When actual cash flow for the period is taking place, it is mon-itored against the projection and any significant variations should be sub-jected to scrutiny.

Once a production programme has been prepared, the contracts manager, site manager, and contract surveyor will discuss and agree the details for a cost, value, and payments analysis as shown in Table 10.12. Costs for indivi-dual months are calculated and also the cumulative totals. The contract is commenced at the beginning of October and at the end of October the value of the work done is certificated. Under normal arrangements this should be

Table 10.11 Consolidated contracts cash flow statement

Contract	July −	July +	August −	August +	September −	September +	October −	October +	November −	November +	December −	December +
45		50		50		50		20		20		30
46		30		30		30		30		20		20
47		20		20		20		20		20		20
48		30		40		40		20		20		20
49		10		10		8		8		8		8
50	—	—		20		20		20		20		20
51	40	—		20		20		20		30		30
52	60		60		20		20			50		50
53	20		20		30			20		20		40
54	30		30		30		30	30		30		30
55	80		80		60		60			100		100
56	30		30		—	—		20		20		30
57				40		40		40	50		50	
58								60	86		123	
59								40	40		60	
60									20		30	
Totals +		140		190		188		208		358		398
Totals −	260		260		180		250		196		263	
Balance at Bank	−120		−70		+8		−42		+162		+135	

Figure 10.2 Cost, value, payments/receipts, Contract 58 (£000's)

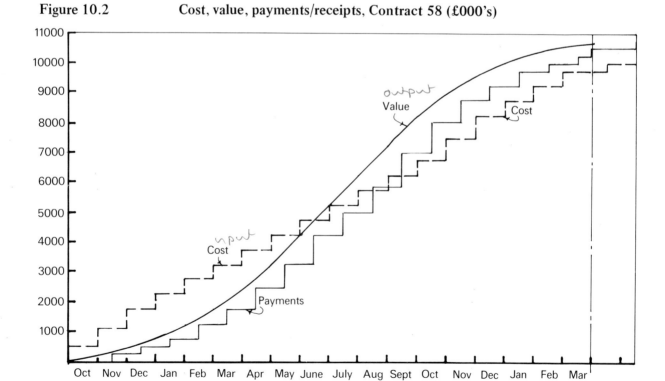

honoured within 14 days. Table 10.12 shows the particulars of contract 58 which was won on a tender sum of £1 070 000. The retention is at the rate of 5% of value with a maximum retention of £25 000. One half of this maximum retention is planned to be released in April, £13 000, the other, in December. Since this is a forecast, calculations are made to the nearest £1 000. Costs are deemed to be those services, materials, fees, etc, which are debited to the contract that month. Value is the value of work done, ignoring any question of retention. One (cost) is input, the other (value) output. The difference between these two is profit or loss at that time. Payments and receipts represent actual cash flow in and out of the firm with regard to contract 58 only. Figure 10.2 shows a graphical treatment of the same information.

Table 10.12   Cost, value, payments/receipts, Contract 58

Month	Cost		Value		Payments/Receipts		Bank	
	Monthly	Cumulative	Monthly	Cumulative	Monthly	Cumulative	—	+
Oct	60	60	25	25	—		60	
Nov	50	110	25	50	24	24	86	
Dec	60	170	30	80	23	47	123	
Jan	50	220	35	115	29	76	144	
Feb	40	260	45	160	33	109	151	
Mar	55	315	80	240	43	152	163	
Apr	45	360	90	330	86	238	122	
May	50	410	95	425	76	314	96	
June	50	460	95	520	90	404	56	
July	45	505	100	620	91	495	10	
Aug	55	560	100	720	100	595		35
Sept	65	625	100	820	100	695		70
Oct	55	680	80	900	100	795		115
Nov	65	745	50	950	80	875		130
Dec	65	810	50	1000	50	925		115
Jan	60	870	25	1025	50	975		105
Feb	50	920	25	1050	25	1000		80
Mar	60	980	20	1070	45	1045		65
Apr	40	1020			(Ret 13)	1057		37
May								
June								
July								
Aug								
Sept	10	1030						27
Oct								
Nov								
Dec						(Ret) 13	1070		40
	(1)	(2)	(3)	(4)	(5)	(6)	(7)	

N.B. The retention sum is 5% of value of work done.
    Maximum retention is £25,000.
    All figures in the table are £000's.

Total work load

When a contract has been won the particulars are entered on a work load form as illustrated in Table 10.13. If the reader examines this table, it will be noted that Contract 46 was commenced in February. The value of the

Table 10.13   Consolidated workload table (£000's)

Contract number	JAN	FEB	MAR	APR	MAY	JUNE	JULY	AUG	SEPT	OCT	NOV	DEC
27	500	400	250	30	—							
28	900	800	600	500	400	200	20	—				
29	700	650	600	550	500	460	420	400	300	200	100	50
30	1100	1000	800	600	400	200	100	50	—			
31	800	400	200	100	50	—						
*32	800	700	1000	800	600	400	200	100	—			
33	950	700	600	500	400	320	220	120	80	—		
34	1000	850	700	600	500	400	300	200	100	20	—	
35	850	600	450	350	250	150	50	—				
36	300	250	200	150	100	50	20	—				
37	100	50	—									
38	400	300	200	150	100	40	—					
39	1000	900	800	700	600	500	400	300	200	100	50	—
40	750	600	500	400	300	200	100	50	—			
41	600	500	400	350	300	250	200	150	100	50	30	—
42	550	500	400	350	320	290	270	230	200	170	140	120
43	220	120	60	—								
44	400	300	200	100	—							
45		2500	2200	2000	1900	1800	1600	1400	1200	1000	800	600
46		1800	1750	1700	1650	1600	1550	1500	1450	1400	1350	1300
47			1500	1450	1400	1350	1300	1250	1200	1150	1100	
48				1950	1850	1800	1700	1600	1400	1200	1000	800
49				800	750	700	650	600	550	500	450	400
50					1250	1150	1100	1050	1000	950	900	850
51						800	750	700	650	600	550	500
52						1400	1300	1200	1000	800	600	400
53						800	700	600	500	400	300	200
54						700	600	500	400	300	200	100
55							2000	1900	1800	1750	1700	1600
56							850	800	750	700	650	600
57								1000	900	850	800	750
58												
59												
60												
Value in hand	11920	13920	11710	14180	13670	15710	16450	15700	13830	12230	10770	9370
Value completed this month	2300	2210	1780	1760	1660	2100	1750	1870				

Time now ——————→

222

contract is £1 800 000. £50 000 worth of work was done in February, reducing the work left to be carried out to £1 750 000. At the commencement of the year contracts 27 to 44 inclusive were in progress, by the end of the year it is expected that all of these, excepting for Contracts 29 and 42 will have been completed. The total of any column is the value of work in hand. To find the value completed in any month, consider the totals in the columns for January and February. The grand total of value of work in hand for February was £13 920 000. From this is subtracted the value of Contracts 45 and 46 just awarded. The value is subtracted from the January total to find the value of work done in January. For example:

	£
February total	13 920 000
Less contracts 45 and 46	4 300 000
Adjusted total for February	9 620 000
January total	11 620 000
Less adjusted total for February	9 620 000
Value of work completed in January	2 300 000

This control can be exercised in the form of projection and a separate control sheet measuring progress against plan. By adding the total amount of work done in each of the first six months of the year it can be ascertained that the firm is carrying out approximately £24 000 000 worth of work per annum. The value of the work in hand at any one time is about six months' work which is, of course, an extremely small amount and would be regarded as very unsatisfactory. Figure 10.3 shows a graphical treatment of the same information.

BUDGETING

An essential part of the strategy of a business enterprise is the identification of the firm's long-term objectives. The firm should know what it is proposing to sell and to what kind of customers. The next stage might be to assess the level at which it proposes to operate, that is, the forecast of the value of the annual turnover.

In order to be competitive, the cost of running the head office must be such that when levied upon the cost of manufacturing the firm's product, the price of the finished goods or services can be compared favourably with those of its competitors. The cost of running the head office will tend to remain fixed irrespective of whether the firm is very busy or slack in manufacturing. In this sense, work carried out on building sites can be equated with manufacturing industry, albeit in a temporary factory. The lower the cost of running the head office relative to the value of work being done on sites, the more competitive the firm. This ratio of fixed cost (head office cost) to variable cost (materials, labour and plant used on site) will vary about many factors but paramount amongst these are:

1. The estimating success ratio. That is, if a firm prepares twenty estimates and is successful in gaining only one job, estimating costs will be high.

2. The contractual conditions associated with the work for which it proposes to tender. For example, a firm might be involved with protracted negotiations for a design and build contract and finally withdraw from the negotiations; this is obviously a very costly process.

3. Some local authorities, because of a desire to be seen to be scrupulously fair, offer work under open tendering arrangements, that is, all and sundry can tender for the work. The committee, charged with the responsibility of making the selection, may feel obligated to accept the lowest tender. A firm with a normal to high overheads percentage to cover fixed costs will tend to do badly in this kind of competition. When there is perfect competition, there is invariably one trader prepared to tender on unreasonably low margin of cover for overheads.

Figure 10.3   **Consolidated workload table graph derived from Table 10.13**

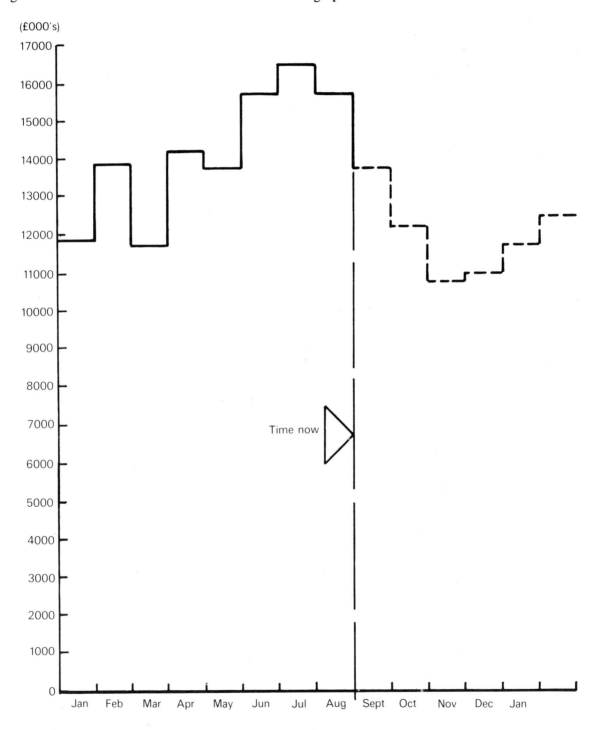

Preparation of a master budget.

It is the responsibility of the Board of Directors to identify the firm's objectives. They will also lay down policy guidelines regarding planned annual turnover and the planned cost of running the head office. The chief executive is charged with the preparation of the master budget to keep running costs within the prescribed limits. It is sound commonsense to involve all departmental managers to participate in dividing up the total planned expenditure between the departments. The proposed expenditure can be sub-divided by departments and also according to type of expenditure. Two broad classifications are capital expenditure and revenue expenditure. If a computer is purchased, the cost of the purchase must be apportioned over a period of years, according to the estimated life of the machine. The rates paid to the local authority in respect of the head office building are an annual charge.

The preparation of the departmental subsidiary budgets and uniting them into an annual master budget is a formidable task. Budgeting is even more complex since no single year can be budgeted in isolation. The purchase of a computer might involve the running out of the present system all pre-computer office machinery, a gradual change over to adapt all machines, stationery, and retrain personnel, to working with the computer.

Participation Committment

The involvement of senior managers in the preparation of their departmental budgets is vital to their motivation. Once they have made predictions in cost terms, they are involved in making the budget plan succeed. The sequencing of expenditure is obviously important to ensure that the resources or services purchased are available when required but not purchased unreasonably before required. The consolidated budget will be compatible in its usage of cash with the cash inflow from contracts or other sources of finance.

Budgetary Control.

It will be necessary in most cases to prepare forward, anticipated quarterly balance sheets, for the coming twelve months.

At the end of each quarterly period, the budgeting officer will analyse all expenditures and allocate them to the departmental cost centres. Each department is a major cost centre. It is then sub-divided into a number of subsidiary cost centres. For example, a Personnel Department would require some or all of the following:

1. Salaries budget
2. Training scheme budget
   (a) Management trainee
   (b) Craft apprentices
   (c) In company training for site managers.
   (d) Institute of Building site managers scheme
3. Construction Industry Training Board, application of levy
4. Recruitment and induction budget
5. Prizes and awards budget

The purpose of a budget is to assist in controlling the direction of a business towards pre-determined objectives. No site manager worthy of the name would entertain the idea of running a site without a master programme. The budget is to the chief executive what the overall programme is to the site manager; it is all about the balanced use of scarce resources. An organisation which has resources idle is inefficient.

Continuous processes

Budgeting and budgetary control are continuous. No sooner is an organisation launched in a new financial year than the executives must be laying detailed plans for the year ahead.

Budgetary control involves the continuous monitoring of performance against plan, actual expenditure against planned expenditure, and the classification of variations from the planned standards into controllable and uncontrollable. Once the variations from planned standards have been categorised, those which are controllable are further classified into significant and insignificant. Executives then deliberate to see what corrective action can be taken concerning the controllable significant adverse variants and what lessons can be learned from the favourable ones.

Action may also be necessary because of uncontrollable adverse variations. If the local authority increase the general rate, a firm will have to pay more rates on the office premises. This is an example of an uncontrollable variation. However, it is necessary to make sure that the extra cash is available to meet the increased cost in rates.

Control limitations

It is almost certain that expenditure will not coincide with the plan. Therefore it is wise to agree upper and lower limits of deviation.

These will be expressed in percentage terms such as plus or minus two percent (Figure 10.4). This gives the budgetary controller the opportunity to operate without constant reference to others. If expenditure is much below that anticipated, it is necessary to know why. At the worst, it may mean that executives are failing to reach targets; this defeats the objective of planning. Targets must be attainable but they should also be realistic. Low targets can be built into a budget if the attainments for one year are increased by some allowance for inflation and become the targets for the next year without due thought and consideration. A fashionable way of dealing with this is to adopt a policy of zero budgeting. That is, the planned expenditure for a year must be judged off considerations for that year and the future, without any reference to past expenditure.

Figure 10.4   **Training budget**

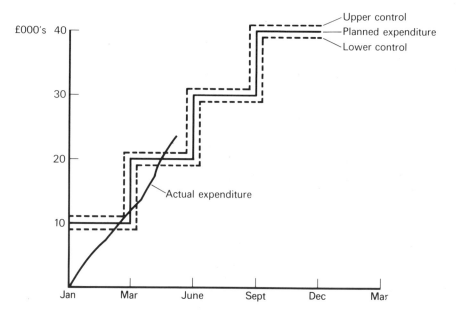

OVERHEADS

The amount of overheads percentage which can be levied upon the direct costs and still remain competitive varies with the type of work, whilst the amount of profit which can be earned varies with the market conditions. General repairs and maintenance require more service from the head office than a speculative housing section. Work being carried out on a contract by nominated sub-contractors requires less head office service than work being carried out by directly employed labour.

Consider a firm engaged on maintenance and repair work with the potential to do £130 000 workload per annum. The fixed overheads are £20 000, which is 15.4% of the total, the profit margin is 7.7% of the total (see Table 10.14).

Table 10.14 **Profit margins**

Line	Level of activity (%)	Workload value (£)	%	Fixed overheads Amount (£)	%	Profit Amount (£)	%	Oncost Total (£)
1	100	130,000	15.4	20,000	7.7	10,000	23.1	30,000
2	90	117,000	17.1	20,000	6.0	7,033	23.1	27,033

Line one of the table shows the objectives. No matter how the level of activity varies, the fixed overheads remain at £20 000. This is true at least in the short-term. The percentage available for overheads and profit is 23.1%. On the reduced workload of £117 000 the fixed overheads rise to 17.1%; this leaves only 6.0% available for profit. This may not appear very serious, but an examination of the percentage drop in profit shows it to be a disastrous 29.66%. $\frac{7\,033}{10\,000} \times \frac{100}{1} = 70.33\%$.

Calculation of overheads as a percentage of the wages of directly employed site workers

Some firms use this method of determining a percentage to build into estimates for the recovery of overheads. This relates the annual cost of directly employed men's wages to the workload. The calculation to find the percentage addition should be done towards the end of one trading period ready for the next.

Example 8.4

	£
Annual work load	130 000
Direct wages	40 000

Direct wages as a percentage of workload = 30.8%.
For practical purposes 31%.

However, if this small firm were normally engaged in projects with an average value of £40 000 and it was noted that the average project value was dropping, the firm would be less likely to attain the £130 000 work load. Smaller than average jobs would probably require the same value in service from the head office as the average job.

In the situation described the following actions might require consideration:

1. Increase marketing activity.

2. Reduce fixed overheads.

3. Reduce the profit margin to increase competitiveness.

4. Diversify the type of work undertaken to widen the market.

5. Assess whether the firm can afford to take no action if there are clear indications of better times in the future.

Overheads related to different systems of employment

Whilst a 20% allowance for fixed overheads may be appropriate for repairs and maintenance work, the allowance for large commercial building work may be as little as 5%. The actual cost in overheads of servicing a site will vary with amongst other things the mode of employment of the workmen (Table 10.15).

Table 10.15

Percentage overheads for different systems of employment

Mode of employment	Overheads percentage for commercial buildings
Directly employed	8
Labour only sub-contractors	6
Nominated sub-contractors	4

Changing the mode of employment

If a firm decide to change from a policy of direct employment to all sub-contracted work, the workload will have to be increased if the same amount of profit is required. Table 10.16 shows necessary variations in the work-level to maintain an equal trading position when differing modes of employment are used.

Table 10.16

Variations in workload

Line	Directly employed %	Domestic sub-contractors %	Nominated sub-contractors %	Work load £	Overheads £
1	100	—	—	1,000,000	80,000
2	—	100	—	1,333,333	80,000
3	—	—	100	2,000,000	80,000
4	60	40	—	1,132,000	80,000
5	30	30	40	1,499,000	80,000

Explanation of calculations for Table 10.16:

Line 2 $\quad \dfrac{\pounds 1\,000\,000}{1} \times \dfrac{8\%}{6\%} = \pounds 1\,333\,330$

Line 3 $\quad \dfrac{\pounds 1\,000\,000}{1} \times \dfrac{8\%}{4\%} = \pounds 2\,000\,000$

Line 4 $\quad \left[\dfrac{\pounds 1\,000\,000}{1} \times \dfrac{60}{100}\right] + \left[\dfrac{\pounds 1\,333\,333}{1} \times \dfrac{40}{100}\right] = \pounds 1\,132\,000$

Line 5 $\quad \left[\dfrac{\pounds 1\,000\,000}{1} \times \dfrac{30}{100}\right] + \left[\dfrac{\pounds 1\,333\,333}{1} \times \dfrac{30}{100}\right] + \left[\dfrac{\pounds 2\,000\,000}{1} \times \dfrac{40}{100}\right]$

$$= \pounds 1\,499\,000$$

Policy

The information portrayed in Table 10.15 and Table 10.16 is not an end in itself. Whether the work is carried out by direct labour or domestic sub-contractor is usually a matter for deeper consideration. Some points which may influence the decision on mode of labour employment are:

1. The contractor may wish to sub-contract a particular risk.

2. For prestige reasons it may be desirable to use all directly employed men on a site.

3. There may be a shortage of men willing to be directly employed or vice versa.

Contribution to overheads

Small building businesses are often divided into sections around the different kinds of work activity. The following divisions are not untypical:

1. Joinery workshop

2. General building repairs and alterations

3. Painting and decorating

4. Plumbing and electrical work.

It is necessary to have a proper system of control in order to know where profit or loss is being created, expenses which are clearly incurred in respect of one section alone are debited to that section. Where an expense is incurred in respect of two or more sections, the cost can be apportioned to them in the ratio of their direct wages bill or alternatively the value of sales which each generates. If one section gives service to the other, it is credited with a sale of that value, the recipient section is debited with the expense (Table 10.17). This would occur if the joinery section made joinery for the general building repairs and alterations section. In order to keep all sections keen and competitive, they may be asked to tender for work awarded to the firm against outside competition. This would only be attempted when the section(s) involved had full order books.

Table 10.17   **The accounts (in abbreviated form) of a firm for a one year period**

		Section of firm			
Item	Joinery	General building repairs and main-tenance	Painting decorating	Plumbing and electrical work	Total
Direct materials	93,000	90,000	14,500	40,500	238,000
Direct labour	43,000	80,000	37,000	30,000	190,000
Services received from other sections	—	20,000	—	—	20,000
Share of fixed overheads	20,000	35,000	17,500	27,500	100,000
Total debits	156,000	225,000	69,000	98,000	548,000
Sales	160,000	235,000	76,000	90,000	561,000
Profit	4,000	10,000	7,000	—	21,000
Loss				8,000	8,000
				Profit	13,000

Analysis of trading results

Some of the senior members of the firm might wish to close down the plumbing and electrical section after analysing the trading results shown in Table 10.17. Approximately forty percent of the profit earned by the other three sections of the business was used to absorb the loss created by the plumbing and electrical section.

However, if the directors did close the plumbing and electrical section they would experience an even nastier shock in the form of a further loss of profit. Table 10.18 shows the revised accounts after the closure of the plumbing and electrical section.

Table 10.18   **The revised accounts**

		Section of firm		
Item	Joinery	General building repairs and main-tenance	Painting and decorating	Total
Direct materials	93,000	90,000	14,500	197,500
Direct labour	43,000	80,000	37,000	160,000
Services received from other sections	—	20,000	—	20,000
Share of fixed overheads	26,875	50,000	23,125	100,000
Total debits	162,875	240,000	74,625	477,500
Sales	160,000	235,000	76,000	471,000
Profit	—	—	1,375	1,375
Loss	2,875	5,000	—	7,875
			Loss	6,500

Closing the plumbing and electrical section changes a trading profit of £13 000 into a loss of £6 500. This is a worsening of the position by £19 500. Whilst the contribution to overheads of a section exceeds the loss created by it, there is a case for keeping the section open, at least in the short-term.

The concept of costing

All costs other than fixed costs are deemed to be variable. When variable costs are deducted from the value of the sales, the remainder is spoken of as the contribution to overheads. Each unit of production must bear its proportion of fixed costs and also the cost of the materials and labour used in its manufacture.

Example 10.5

The fixed costs of running a small furniture-making factory are £20 000 per annum. The timber used to manufacture one chair cost £5.00 and the labour £3.00. Determine the unit cost of manufacturing 1 chair, 99 chairs, and 999 chairs respectively.

Table 10.19

Number of chairs	1	99	999
Fixed costs (£)	20,000	20,000	20,000
Material (£)	5	495	4,995
Labour (£)	3	297	2,997
Total (£)	20,008	20,792	27,982
Unit cost (£)	20,008.00	210.02	27.99

Derivation of control ratios

It is assumed that the fixed costs are made up as follows:

	£
Factory overheads	14 000
Head office overheads	4 000
Marketing and sales cost	2 000
	20 000

If the chair manufacturer, after examining the market, his production capacity and working capital, decides that there is the opportunity to make and sell one thousand chairs per annum, he can relate his performance in the future to the following control ratios:

1. The standard absorption rate

$$\frac{\text{All overheads}}{\text{Standard production}} = \frac{£20\ 000}{1\ 000\ \text{chairs}} = £20.00 \text{ per chair}$$

2. Factory overheads ratio

$$\frac{\text{Factory overheads}}{\text{Standard production}} = \frac{£14\ 000}{1\ 000\ \text{chairs}} = £14.00 \text{ per chair}$$

3. Head office ratio

$$\frac{\text{Head office overheads}}{\text{Standard production}} = \frac{£4\ 000}{1\ 000\ \text{chairs}} = £4.00 \text{ per chair}$$

4. Market and sales ratio

$$\frac{\text{Market and sales overheads}}{\text{Standard production}} = \frac{£2\ 000}{1\ 000\ \text{chairs}} = £2.00 \text{ per chair}$$

After one year of producing chairs, the manufacturer extracts the following figures:

	£
Factory overheads	12 000
Head office overheads	5 000
Marketing and sales cost	4 000
	21 000

These control ratios now become:

1. The standard absorption rate

$$\frac{£21\ 000}{1\ 000\ \text{chairs}} = £21.00 \text{ per chair. This is an unfavourable ratio.}$$

2. Factory overheads ratio

$$\frac{£12\ 000}{1\ 000\ \text{chairs}} = £12.00 \text{ per chair. This is a favourable ratio.}$$

3. Market and sales ratio

$$\frac{£4\ 000}{1\ 000\ \text{chairs}} = £4.00 \text{ per chair. This is an unfavourable ratio.}$$

It should be noted that ratios point towards problems but do not reveal solutions.

Use of the ratios

These are just a few of the many ratios which are in use in manufacturing industry. Similar ratios can be developed, for example, for speculative housing developments.

When a developer occupies a site, many costs are incurred which are not directly proportionate to the number of houses being built. Amongst these are:

1. The cost of site hutments.

2. Making access to the site.

3. The cost of temporary services.

4. The cost of bringing permanent services on to the site.

5. The cost of providing site management.

Whilst it is conceded that the ratios demonstrated cannot be applied with the same precision to these building site expenses, not dissimilar calculations must be made to determine the optimum number of units that a builder can erect on a single site to maximise profit.

THE COST CONTROL OF BUILDING SITE WORK

The basic principle underlying a cost control system is to find the cost of work whilst it is proceeding and compare this to planned standards. Where the comparison is unfavourable, enquiries should be made and, if necessary, corrective action taken.

Advocates of costing systems point out that since control information highlights the weak or potentially weak areas, managers can use their time and talents where the rewards are greatest, the application of the "management by exception" principle.

To be of maximum value, costing information must be provided at regular and frequent intervals. The manner of presentation should allow easy comparison with planned standards. Cumulative total comparisons as well as weekly comparisons are important since they reveal the underlying trends. One generally expects productivity to be lower at the commencement of a new operation than when the method of working is established and understood. This 'working in' process is usually referred to as the learning curve.

Some advantages claimed for costing systems

1. It ensures that first line managers become more involved in planning work and measuring performance.

2. It reveals to senior management the strength and limitations of site managers, thus helping to select the right man for the job and it also acts as a guide when formulating training programmes for site managers.

3. Design weaknesses may be revealed and hopefully the designer may modify the design to improve the buildability factor.

4. Profit or loss on operations can be known whilst the work is progressing.

5. It is particularly useful when trying out new methods of working or new items of plant in limited area trials.

6. The effect on productivity of different wage payment systems and different qualities of labour can be measured.

Figure 10.5   **Cycle of efficiency**

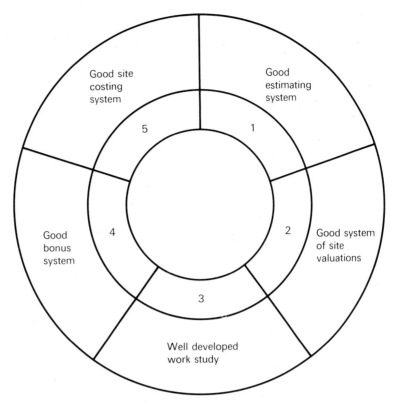

Some disadvantages attributed to costing systems

1. It is yet another expensive control system which produces no wealth.

2. There are so many variables which might upset production, such as failure of the designers to provide the necessary information, subcontractor performance failure, material supply failure; there is little point in measuring what is bound to go wrong.

3. Cost control is looking backwards, the site managers should be looking forward and getting on with the job. Good site managers can assess what is a fair performance.

4. Costing systems, to be of value, require that accurate records are maintained of the hours input to each separate task. Only those who are on the spot know the input but concrete gangers, navvy gangers, etc, are not conspicuous amongst the best recorders.

5. The system is checking on the performance of people, it mitigates against building up a good team spirit atmosphere.

6. Different architects may present a contractor with the same or similar specification but their separate interpretations of the specification may be completely different.

7. Bill rates represent an average value for an item of work but outputs on the same work item can reasonably vary quite a lot. For example, the output of a bricklayer below window sill level on the long run of brickwork will be much greater than when he builds the short runs between the window openings.

Table 10.20   **Weekly costing sheet for brickwork**

	Operation	Planned output rate per hour	Total quantity	Total hours allowed	This week					Cumulative				
					Measure	Hours	Actual output rate	Hours gained	Hours lost	Measure	Hours	Actual output rate	Hours gained	Hours lost
1	1B wall in foundations	0·5 m²	200 m²	400	60 m²	150	0·4	–	30	180 m³	430	0·42	–	70
2	½ B facings outer skin of cavity wall	0·75 m²	150 m²	200	60 m²	75	0·8	5	–	130 m³	150	0·87	23·3	–
3	75 mm Clinker concrete block walling	3·0 m²	500 m²	167	180 m²	70	2·57		10	400 m³	120	3·33		13·3

5   –   40

Hours lost = 35

23·3 – 83·3

Hours lost = 60

Operation 1   Rate of loss this week is adverse compared with earlier production

Operation 2   Considerable improvement this week

Operation 3   Considerable deterioration in output this week. Previous to this week production total was 220 m³ in 50 hours or a rate of 4·40 m³ per hour

Some of the factors to be considered when setting up a costing system

Management efficiency

Costing systems are likely to be successful only if the management of the firm is efficient. If there is a cycle of events, it is probably as depicted in Figure 10.5.

Sequence

Each of the five sectors is about measuring work and as a developing firm becomes competent, it moves from one sector to the next. The total should be an integrated system permitting information collection with the minimum amount of duplication.

Coding

A comprehensive coding system must be prepared so that any item of work which is billed separately can be identified by a unique numbering system.

For example:

B1 ½B wall in common bricks mortar.

B2 ½B wall in common bricks, one face jointed as the work proceeds.

B3 ½B wall in common bricks, both faces jointed as the work proceeds.

Communication

All those involved with the system should know how it works.

Control

Daily labour allocations should be completed daily by the first line supervisors and checked not later than the following morning by one of the Site Manager's staff whilst everything is fresh in everyone's mind.

Speed

Costing systems should produce control information speedily so that when applicable corrective action may be taken whilst the operation is still in progress.

Labour

Plans should be made at the estimating stage on which trades will be sublet and also on the wages and bonus policy towards directly employed men.

Materials

Consideration must be given in the system to accounting for materials received but not yet invoiced. User waste should be measured and analysed.

Table 10.21   Calculation of waste form

a Material	b Qt. del vd. to date	c Qt. in Valuation	d Materials on site	e c & d	f b - c	g Apparent waste e - f	h Waste allowed	j Excess waste	k Costs	l Remarks & remedial action
Facing Brick	10 000	4 000	4 000	8 000	6 000	2000	500	1500	£90.00	Alter stacking arrangements.

Table 10.22   Weekly plant cost sheet

WEEKLY PLANT RETURN							CONTRACT NO: Date:
		COMPANY OWNED PLANT					Completed by:

Register number	Rate	Hours				Charge to contracts	Remarks
		Worked	Idle	Broken down	Maintenance		
CM13	£20.00 p.w.	32	4	2	2	20.00	Broken down time charged to contract-site neglect
CM48	£30.00 p.w.	28	8	4	—	27.00	Faulty bearing to drive shaft
Ex7	£8.00 p.h.	32	6	—	2	320.00	Check work planning to reduce idle time
Ex11	£8.00 p.d	30	4	3	3	38.50	50/50 allocation of broken down time-questionable whether filter replaced correctly at last routine service

Plant

Plant which is engaged during the week on many different items of work which are to be allocated to different cost centres presents a problem. The best solution may be to check it weekly against the sum total of hours allowed for plant in those items rather than an allocation to cost centres.

Table 10.20 shows part of a weekly labour allocation sheet for bricklayers.

Table 10.21 shows a weekly material waste schedule.

Table 10.22 shows a weekly plant summary sheet.

11. Marketing

Introduction

"Marketing is not just selling; selling is part of marketing. Advertising and promotion are not marketing; they are part of marketing. Marketing is not just market research; research is also part of marketing and, indeed, I believe that manufacture is also part of marketing. Marketing is a perspective, a way of looking at the operations of a business, a way of doing business. A marketing perspective puts the customer at the centre of a firm's activities and orientates the firm towards its markets and customers rather than towards its factory and production lines."

Leslie Rodger

Marketing the product is the basis of industrial activity and all other aspects of managerial activity are allied to it. Marketing, production and finance are the three anchors to which all other activities are linked. Yet in

Figure 11.1   **Interrelation for most business activity**

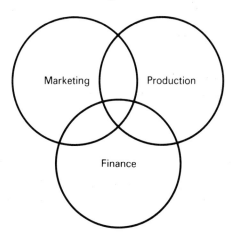

the study 'Efficiency in the Building Industry' (1) ample evidence is available that the participating firms gave great consideration to production and financial matters, but to quote: "Marketing played little part in the activities of the firms".

If it were possible to measure the improvement in efficiency of contracting firms over the past twenty years in the following aspects, there would be little doubt about which function has been most overlooked.

1. Production techniques and control.

2. Financial management.

3. Personnel and industrial relations.

4. General administration.

5. Marketing.

The demand for building services

Central government and local authorities are responsible for commissioning more than one half of the construction activity in the United Kingdom. However, in the civil engineering sector, the public purse is responsible for commissioning around ninety per cent of the work available. Thus a change in policy by central government regarding the level of expenditure on roads, bridges, reservoirs, etc, has a very direct effect on this kind of activity. The fact that governments are elected for relatively short periods probably does not improve the prospects for good long-term planning and continuity of work.

Interest rates

Most clients for large projects, both public and private, borrow money to finance building activity, and therefore interest rates become central to the level of building activity for the main contracting section of the industry. In prosperous times industries invest in new plant and buildings, and employed personnel seek to buy new houses; thus the demand for new work in the private sector is allied to national prosperity.

Maintenance

About twenty-five per cent of the work done by the industry is maintenance, which includes painting, decorating, alterations, refurbishing and small works generally. The availability of maintenance work tends to be fairly constant, despite the changes in national prosperity.

Overseas work

The value of work carried out by British contractors overseas is now around twenty-five per cent compared to the value of work done in the United Kingdom.

Building materials

In 1978 the building materials industry of the United Kingdom exported £1 000 000 000 worth of materials and components.

What kind of market

Housing

In the private sector the main demand is for two-storey housing in the lower price range. The relatively high cost of land and high interest rates encourage quick construction and quick sale. There has therefore been a strong move towards timber-framed housing, and this, coupled with internal dry lining systems, has reduced construction time. The influence of the House Builders Registration Council is now having the effect of encouraging house-builders to improve their marketing expertise and to meet certain standards of quality control. Without the Council's certificate a would-be purchaser is unlikely to obtain a mortgage from the main building societies.

In recent years there has been a noticeable increase in the number of houses built for housing associations, and this has meant that in many instances the house-builder in the private sector (building houses at the lower end of the price range) is brought into the situation where he works under the control of an architect. The customer knows what he wants, whereas normally the customer is offered a product by the builder and he either takes it or leaves it.

There is ample evidence to show that there are many people renting homes who could afford to buy. There is also a good deal of ignorance amongst people who have never purchased a house about mortgage arrangements and legal matters concerning house purchase. There are few products which one can guarantee will sell for more than the purchase price twenty years on.

Competition

Smaller firms naturally have a more restricted geographical area of operation, although improved road systems have made it possible for some to operate over greater distances. Against this improvement in communications must be weighed the fact that in times of depression the cost of travelling reduces thin profit margins even further. Although it would appear at first sight that firms tender for the size of job according to the capital formation, in a depressed market many firms naturally move downwards and try to obtain contracts which in more normal trading conditions they would consider too small. The effect of this is to make survival for the smaller contracting firm very precarious.

Obtaining work

There can be few other industries, if any, which employ such a small ratio of their personnel as specialists in marketing. Obtaining work is so often the outcome of a favourable recommendation and it is likely that this will continue to be the way by which much work is obtained for a long time to come. Obviously, if a client has previous satisfactory experience of a building contractor, he is going to employ the firm he knows and respects. This highlights the need to stress to all staff coming into personal contact with the client the importance of their marketing role.

Small projects

Some 'Yellow Pages' classified telephone directories list more than seven hundred building contractors. Most of these are single line entries. Very few mention their affiliation to a trade protection society such as the National Federation of Building Trades Employers or the Federation of Master Builders, nor is there any indication as to whether the proprietor is technically qualified. If the problem were to select a garage, after extracting a name from the Yellow Pages one could visit the establishment and make an overall assessment. However, this is not possible with the builder, the argument generally advanced being that the prospective client has no way of determining from the entry what particular qualifications the builder holds.

One way in which the smaller firm may advertise is by having a well prepared brochure. Well prepared means prepared by a graphic designer. Ideally, a firm should have more than one brochure so that they can identify with a wider range of projects. The owners of properties obviously needing a builder, either on account of dilapidation, road improvement schemes or parking and garaging problems, and those eligible to take advantage of development grants might have a brochure posted to them and a follow-up call a few days later. Every effort should be made to check the suitability of the client before marketing. The person making the face to face follow-up should be sensitive to the state known as pre purchase anxiety. It is essential in the follow-up stage to be addressing the head of the household. It is more productive to scatter the marketing described two houses here, two elsewhere. Once successful, some work will follow by recommendation if the job is priced sensibly and carried out properly.

There is nothing unethical in the seller seeking out the buyer.

The larger job

There are in existence private organisations which obtain all the planning applications being submitted to each authority in a region and publish these in lists, making these lists available to contractors for a fee. This gives the contractor a starting point for making further enquiries.

If a firm wishes to obtain information about potential jobs before the professional investigators have published their list, it will be necessary for members of the firm to visit local libraries or local authority offices to consult the planning applications list as soon as this information is available. Whether it is possible for one person to obtain this information, or whether it requires the involvement of a number of senior personnel depends very much on the number of planning authorities involved and their proximity to each other.

Choosing a contractor

The Royal Institute of British Architects operate, through their RIBA Services Ltd, the Product Data service. This is a library system for architects and is regularly updated. In conjunction with this service, they circulate a brief description of subscribing building contracting firms. On an A4 sheet is given a general description of the firm to give an indication of its size, type of work, and supporting services by way of special departments or subsidiary companies. A few photos of completed jobs may be included. On the reverse side a list of recently completed projects, together with the name of the client, architect, the approximate value of the contract and the contract duration, is given.

Use of the press

Buying a small amount of space on one side of the title block at intervals of once a month in the local or national press ensures that the contractor's name is easily recalled by the reader when he is thinking about building firms. Obviously, everything that is newsworthy and likely to improve the image of the firm should be mentioned in the press. There are, of course, the usual activities such as cutting the first turf, laying the foundation stone and planting the first tree. If it is possible to think of more original activities, the advertising 'pull' may be stronger.

On-site publicity

Well designed site boards carefully located are of considerable value in bringing the firm's name to the attention of the public. It is, however, important that the public suffer the minimum of inconvenience from the project, otherwise the name becomes synonymous with creators of public inconvenience. Since others (sub-contractors and design specialists) will be advertising from the main contractor's board, the revenue from these sources should offset the major costs. It is important to keep site boards clean, otherwise much of the value is lost. Where circumstances permit, 'side walk directors' viewing platforms can be provided. Apart from the publicity aspect, these platforms can relieve congestion problems caused by members of the public obstructing the access points. One does not have to be a builder to recognise a tidy site and a businesslike atmosphere and it is important to pick the right team for the key sites.

Housing

On the housing site there are similar opportunities. First and foremost, it must be remembered that in the case of private housing the client is the user. In order that he can better appreciate what he is considering buying, a show house is certainly a necessity. Where the scale of operations is large enough, the show house should be manned for the weekend, when husband and wife are free to view together.

The public expectation	People do not expect to be laying lawns and paths as they might have done in the immediate post-war years. The builder who provides these services and who also clears up before houses are occupied is more likely to create customers than one who does not. When the site is partially occupied but still under construction, separating the building operations from the occupied premises will be much appreciated and generate goodwill. Also, making absolutely sure that the garden contains garden soil and not rubbish or sub-soil is essential.

Conclusion

Enhancement of the firm's reputation

Dealing professionally with the present client creates the next customer. The following points summarise the factors which are likely to impress the smaller private client:

1. Prompt attention to enquiries.

2. One member of staff seeing the job through as the firm's sole representative from enquiry to after sales check-up.

3. A fully itemised and well-written estimate, written in language that the reader will understand.

4. In the absence of an architect, a high quality of advice concerning the proposed work. This might include arranging a visit for the client to view something of a similar construction to that suggested.

5. Advice and assistance with regard to financial arrangements.

6. If alterations to the original agreement are made during the course of the work, a frank discussion on the cost implications at the time when the matter arises.

7. Guarantees of quality and maintenance arrangements after completion.

8. Instructing the client in the operation of services and equipment provided.

9. Absolute compliance with the specification and performance standards.

10. Maintaining a high standard of cleanliness and tidiness during the building operation.

11. Making the minimum disruption to the client in his domestic or commercial operations. This might include arranging car parking off the client's premises for the client's vehicles, the vehicles of the building operatives and company owned vehicles.

12. Keeping to the original agreed construction programme.

13. Getting off the site promptly on completion.

14. Promptly presenting a final account.

15. Making a personal call at the client's premises some months after completion of the work to ensure that all is satisfactory.

People in marketing

It seems to be one of the facts of life that people entering the industry and being trained through the normal technical channels often do not recognise the importance of marketing. The industry is also an all male society. Compare the approach of a busy site manager to a casual site visitor with that of a motor car salesman and the need for a construction sales force lurking in the wings seems desirable. There seems little doubt that constructors will have to pay much more attention to employing sales staff in the future.

BIBLIOGRAPHY

1 Efficiency and growth in the building industry. Building Economics Research Group of Ashridge, Management Research Unit, Eleanor Lea, Peter Lansley, Paul Spencer.

12. Maintenance Management

The significance of the maintenance sector

All the foregoing chapters have been based upon the assumption that the work which is being done is new, or at least a substantial extension to existing property. In reality much building work is not concerned with the erection of new buildings but rather with maintenance or renovation. The current policy of augmenting the programme of new housing by a considerable amount of rehabilitation of old housing stock adds a further load to this sector. Of the total building labour force employed in the UK, over 40% are employed full-time on maintenance and minor works, and in a period of a declining volume of new work it could well be that this percentage has already increased. The national annual expenditure in this sector together with engineering services is estimated at over £3000 m.

It is therefore important that some consideration should be given to this significant sector of the industry and to the particular management problems which are generated by the characteristics of this type of work where pre-planning is much more difficult and the extent of a repair much less well circumscribed.

Maintenance management may be viewed either:

(a) from the viewpoint of the property owner who needs to employ a manager to maintain his estates in good working order, or

(b) from the viewpoint of the contractor who is carrying out the work, who may be part of the property owner's organisation or an independent firm.

Both these aspects need to be covered in this chapter as they are to some extent interrelated. However, it is not the authors' intention to offer more than an introductory chapter on this subject and certainly not to provide a guide to estate or property management which is already covered by extensive literature; a selective reading list is included for those who might wish to pursue the subject further.

Thus this chapter concentrates on the organisation of maintenance and minor works and their execution from the two viewpoints mentioned above, giving a greater concentration upon the actual execution of the works than upon the aids to policy making for the estate manager.

Nature of works

The compass of maintenance and minor works is concerned with a multiplicity of tasks and skills which demand a breadth of experience and technical expertise not necessarily found in new work. They call upon an ability to assess each problem individually, to recognise the symptoms of failure and thus ascertain its cause, and to conform to the standards which each job requires.

In recognising that maintenance may also include an element of improvement and minor works, the definition given by the Department of Environment Committee on Building Maintenance in its report of 1972 sets out clearly the nature and objectives of this work:

"Work undertaken in order to keep, restore or improve every facility, ie, every part of a building, its services and surrounds, to a currently accepted standard, and to sustain the utility and value of the facility."

DOE Committee on Building Maintenance, 1972

The concept of "a currently accepted standard" assumes that standards may be expected to improve with the passage of time and that users' requirements will often demand higher performance levels for their buildings than were the norm when the buildings were designed. Simple examples of this are the upgrading of plumbing facilities in housing in many urban improvement schemes and the provision of central heating and insulation as essential features of modern dwellings. In some cases the improved standard may have become a statutory obligation, as in the case of thermal insulation, whilst in others, as central heating, it has become a socially accepted comfort requirement. This definition also assumes that standards can be identified and described sufficiently clearly for both client and contractor to agree upon an acceptable level of workmanship and performance. This is an important aspect of the management function and is therefore considered in some detail later in this chapter.

Maintenance work

Maintenance work generally falls into two categories in that it may form part of an organised and planned programme of renovation, improvement and servicing, or it may be occasioned by the sudden failure in performance of some part of a building or its equipment. In the first instance the need is foreseen and the work will normally be scheduled well in advance as a part of an overall scheme of maintenance management, based upon past records of building performance and the extent of a particular job will normally be quite well defined. In the second case the occurrence is unforeseeable and the extent of the job is more liable to vary, and is often not ascertainable until after preliminary investigation or emergency work has taken place.

It is clear that work based upon a programme of improvement will fall into the first category though an unexpected demand for repair work due to breakdown or damage may well create the opportunity for bringing forward some part of the longer-term programme.

Types of maintenance work

The method of dealing with work depends upon its nature and before studying this in more detail it is necessary to classify the types of work likely to be encountered which fall into the following broad categories:

1. Preventive maintenance: work carried out in anticipation of failure usually on a planned cycle related to the long term preservation of the building and unlikely to have a high degree of urgency. Cases include redecoration programmes, replacement of roofs, floors and engineering installations.

2. Day-to-day repairs: usually requested by the occupier, and consisting, for the most part, of small jobs which are regarded as urgent in the eyes of the occupier. Prompt attention may avoid more extensive work becoming necessary. Examples may include minor water leaks, glass replacement and easing of doors.

3. Emergency repairs: those which have already caused, or if left unattended will cause, some failure in performance, involving some degree of hazard. Examples are:
   (a) loss of a facility, eg, electricity supply, gas or heating system
   (b) interruption to production or commercial activities, eg machine breakdown or deterioration in environmental conditions
   (c) security risk due to such causes as broken windows or doors
   (d) damage likely to result in more extensive damage and more costly repairs due, for example, to a leak in a service pipe, or extensive roof damage
   (e) structural failures likely to cause injury to the occupier or the public at large.

4. Servicing work: usually carried out by specialists either under contract or direct commission. Covering such items as lifts, refrigeration, water treatment, fire and burglar alarm equipment, heating and ventilating systems.

5. Minor new works: these would consist of works of improvement or modifications to meet the requirements of new legislation. Such work is mostly the subject of a single specification and contract let on tender, thus constituting a series of one-off jobs each essentially the same yet each entirely different.

The role of the maintenance manager

With the newly awakened awareness of the importance of maintenance, the professional nature of the maintenance manager's function is becoming increasingly recognised. It requires not only a sound knowledge of building construction and experience in applying it to the solution of maintenance problems, but also an ability to manage the planning, costing and execution of the works using the most modern techniques.

At present most maintenance managers tend to be essentially practical men with little time for theorising. Yet theory and practice are complementary and it is necessary to aim for a balanced mixture of the two in the basic professional disciplines.

As the maintenance load increases, the planning of the work will pose difficult and exacting problems of technical management requiring analytical skills of a high order which will test the professional competence of any manager. Degree courses to train students in this field are now well established and many graduates are now emerging equipped with the necessary analytical skills to supplement those of the older school whose expertise is based upon long personal experience. It is hoped that this development will do much to raise the status of the maintenance manager in the various employing organisations so that they may be given an opportunity to demonstrate the essential role which they have to play in the well-being and effectiveness of the organisation as a whole.

Theory provides a structure for relating different types of knowledge so that appropriate criteria can be devised for developing and testing new techniques. The graduate will be able to draw upon his knowledge of maintenance technology, management techniques of organisation and control, and industrial sociology, to impart new thinking into the integration of the variety of services with which he will be concerned. It is not unlikely that many organisations in due course will see fit to employ professional consultants to advise on the planning and costing of maintenance work and to oversee its execution on a basis of continuing employment.

The maintenance manager is the man who has ultimate responsibility to the Board of Directors for the whole building estate of an organisation. He is therefore responsible for the formulation of the maintenance policy and also for deciding how it will be executed.

To formulate a comprehensive and effective policy requires consideration of both the short and long-term technical needs of the building being maintained and of their effect on the activities which are housed. He must be aware of the technical, legal and social implications of his policy and their connection with the operations of the whole organisation, and must be able to discuss the cost and human benefits of his proposals with the Board.

It has already been suggested that he has a fully professional role to play and as such must be integrated into the management hierarchy of the company and not considered as an appendage carrying out a necessary but rather subsidiary job of work, out of the mainstream of activities. It is most important that he has responsibility for maintenance in all its aspects and for the budget which relates to the true cost of occupying and maintaining the estate as this is often subdivided between several accounting centres. Indeed it is only when the Board is fully aware of the total cost involved that the maintenance manager will receive the recognition necessary for him to carry out his function effectively.

The maintenance manager's detailed function is mainly of a technical nature and concerned with the planning and cost of construction resources to ensure that the necessary repairs and renewals are carried out as effectively as possible, though his role as an adviser on building performance is leading to his more frequent participation in other spheres of an organisation's activities. For example, it has been said often that maintenance starts on the drawing board and his advice at this stage can aid considerably such matters as the integration of fabric and service installations, and in the selection of materials for specific functions.

However, in the main the manager's major decisions relate to:

1. The determination of standards.

2. The planning of inspection routines.

3. The identification and specification of work items.

4. The estimation of the cost of the work.

5. The planning of the work.

6. The organisation of its execution.

Each of these functions should now be considered in more detail.

Standards

The maintenance manager must ensure that the level of standards equates to the objectives of the organisation whilst meeting statutory and other external requirements. One of his most exacting tasks will be to recognise the varying degrees of disrepair, for example, at which user activities may become seriously affected and the point at which the building loses visual acceptability. The level at which standards are set can greatly influence not only the cost of the work for the client but also the manner in which the contractor will approach the work and allocate his resources.

"Acceptable standards" must be capable of simple definition and three general categories can be identified, as follows:

1. Functional performance, quality and reliability related to user needs.

2. Safety aspects of structure, electricity, fire, etc.

3. Preservation of the asset and its amenities in terms of the interests of the owner.

In each category, however, it is possible to identify various levels of excellence not only for different buildings but equally well for different rooms in the same building. This could be shown as follows:

Level 1 — very high — board rooms, operating theatres.
Level 2 — high — entrance areas, hospital wards, restaurants.
Level 3 — average — offices, dwellings, schools.
Level 4 — low — store rooms, warehouses, limited life property.
Level 5 — very low — prior to demolition.

In reality the higher limit is set by the cost of achieving it, whilst lowering the limit increases the probability of failure thereby involving greater repair costs and other losses in terms of output or comfort standards.

It is most desirable that some form of quantified classification should be devised to define the grades of standard and relate to the conditions which may be associated with each, thereby setting certain guidelines to the appropriate level of expenditure which can be justified for maintenance work.

The factors which influence the determinants of maintenance standards are complex and are an amalgam of economic considerations and the provision of adequate amenity. Much work is being carried out on deciding rules for levels of acceptability and the standardisation of specifications and descriptions in order to build up a body of information not only to guide the designer in setting the appropriate standard but also the maintenance manager in advising the client on the cost of its upkeep.

Inspection routines

The frequency and nature of inspections must depend upon the characteristics and rates of deterioration of building elements so that defects may be identified before they have reached the stage where they are liable to fail.

A sound knowledge of the performance of building elements under a variety of user and environmental conditions is necessary as the guide for planning the inspection routines for specific buildings. Such inspections must be based upon information sheets relating to the building's condition, its past and present usage, defects experienced, causes and symptoms, mechanical and other services, and environmental conditions.

Inspections are undertaken for a number of different purposes and the information required will vary not only in nature but also in depth of detail. For example, an inspection may be carried out:

1. To record the present condition of a fabric and its facilities and give some indication of future maintenance commitments.

2. To detect incipient faults and defects, and the extent of remedial work necessary to restore to the agreed standard and avoid the defect recurring, and to submit recommendations on the degree of urgency for the work.

3. To check that previous work had been carried out satisfactorily and had proved adequate to meet the specific conditions and recommend on the frequency of future inspection.

It is most important that inspections are thoroughly pre-planned. The provision of comprehensive check lists will give some structure to the inspection routines and ensure that all the parts of the building are covered. The procedure for an inspection should follow these stages:

1. The form and purpose of the inspection.

2. The frequency pattern.

3. Qualifications and status of inspector, whether tradesman, technician or specialist.

4. Reporting procedures for feedback of information.

5. The preparation of proformae for each element and sub-element providing:
   (a) description of element and its location coding,
   (b) criteria of performance needed by the inspector,
   (c) space for the inspector's report under each heading,
   (d) recommended remedial work,
   (e) comments on the cause, whether identified or suggested,
   (f) priority rating for remedial work.

Identification and specification of work items

This forms an integral part of the inspection sequence, interpreting the information which has been collected and relating it to the established standards. In the case of a proforma being used on the lines suggested above, the inspector may well carry out this function himself and the maintenance manager's role may be one of checking and confirming these entries from his experience and training, which will have equipped him with the knowledge of the causes of defects over a wide spectrum of situations.

The feedback procedure is of great value at this stage, as it should offer a channel of information back to the designer so that similar defects will not recur in new property commissioned by the organisation.

Estimation of cost

Much has been written elsewhere in this book regarding tendering and estimating methods, but some additional consideration must be given to those aspects particular to maintenance work.

Maintenance managers are usually required to predict costs for both long-term and short-term programmes and these will be based as far as possible upon records of similar work carried out in the past, with any necessary allowances to cover variations resulting from inflation, building types or conditions.

Long-term estimates

These may relate to programmes covering a number of years and groups of properties or individual buildings. The problem is that the precise nature of the work which is likely to be entailed cannot be predicted with any certainty and the estimate must be based upon the average cost of maintenance over past years related to some parameter of the property included.

Parameters which are used for obtaining quick estimates are:

1. Costs per unit of accommodation
   This parameter relates through the units chosen to the number of people using the building and gives a rapid calculation of the likely maintenance expenditure for a number of buildings of similar type. The

weakness is that it overlooks any particular features to be encountered in individual buildings.

2. Costs per unit or floor area or volume

These provide quick and simple methods of obtaining a rough estimate of cost and can be separated into areas requiring different levels of maintenance expenditure. The visit costs are obtained from past records, suitably updated and adjusted for constructional form, exposure, etc.

3. Costs per building element

An analysis of past maintenance costs of individual building elements, ie, external and internal walls, floors, roofs, etc, should indicate a pattern of expenditure related to each element. Whilst these costs could well be cyclic and vary in pattern between the usage and life expectancy of elements, average costs can be determined over several years and should also indicate periods for planned maintenance. If a sharp rise in repair costs occurs for an element it will draw attention to this fact, calling for an investigation to find out the cause.

The maintenance characteristics of a building are determined largely by the initial design decisions and a deliberate maintenance policy should be planned into the design. It has been suggested that the designer should be made aware of the "maintenance profile" for the building through the client's maintenance manager or his own specialist maintenance consultant. This profile would consist of the maintenance characteristics in words, figures or graphical means which can form the basis for the long-term management, financial or technical policies of the maintenance manager who must, therefore, accept the responsibility for drafting this profile. This concept was first outlined in a paper by K Kerward given at the 3rd National Building Maintenance Conference in 1971 which gives further information on this approach to long-term estimating and planning.

Medium and short-term estimates

Estimating in these contexts follows very closely the patterns covered in another chapter of this book which relates to programming, scheduling and controlling the execution of building work. Several methods have been used, the main three being as follows:

1. Analysis of past records

A reasonable indication of the labour and material content of simple jobs can be obtained from past records of similar works, providing data which can be priced at the current rates. Larger jobs may also be broken down into a series of discrete parts, ideally referring to a sequence of trade operations.

It must be accepted that the estimates obtained in this manner will only be approximate and cannot reflect the characteristics of an individual job. The data upon which they are based are taken only from records of cost which give no indication of either influencing conditions peculiar to individual jobs, productivity levels obtained, unproductive time or lack of uniformity in job descriptions. As the volume of feedback increases, so will be the range covered and a more comprehensive data bank can be compiled; however, this will still suffer from many of these disadvantages.

2. Experience

Many tasks are of the 'one-off' kind where past records are not available for comparison and only the experienced judgement of the foreman in charge of the work may be used for estimating. This will normally take the form of a total cost based upon the labour required and the time needed for completing the work. This method is both quick and often surprisingly accurate but inconsistencies between foremen and the lack of formal feedback procedures for checking make it a far from ideal method.

3. Bracketting

Small non-repetitive jobs may be dealt with by grouping them within time brackets by reference to common jobs. Thus a series of jobs may be grouped under headings of the average time per job in batches covering 8 hours or a typical working day.

From the frequency and average time of jobs under each heading the total time and labour costs for a period can be calculated. The cost of materials can be expressed as a percentage of the total labour cost, reflecting the broad nature of the work being undertaken.

This method is suitable for estimating work where the exact nature is not known but which nevertheless may follow a regular statistical pattern, as in the case of much contingency maintenance.

4. Work study data

Whilst this is an expensive method of obtaining information, the accuracy of the data may justify its use in those cases where a well defined statistical pattern of work can be recognised or a limited range of tasks is envisaged.

The importance of the estimating function will differ in cases of the employment of direct labour or of an exernal contractor. In the first instance the basic data will also be used as a foundation for a production control procedure, whilst in the latter its primary use is to predict the amount for which the work is likely to be done. These two forms of executing maintenance work are further considered below.

The planning of the work

The work load of maintenance comprises the sudden emergency requests from occupiers for prompt action in repairing failures and the more substantial long-term programme of repair and maintenance work. The maintenance manager's function is to plan the work so as to obtain a balance between these two demands so that essential work can be executed with the maximum economy and minimum inconvenience to the occupants. Simply stated he must determine the starting and completion dates of the work, bearing in mind its effects upon user activities, its level of priority, the availability of physical and financial resources and the duration of each constituent operation.

Planning approaches generally divide into two complementary and interacting systems:

1. *A scheduled or predetermined system* where the inspection procedure and remedial work are carried out at predetermined times as closely related as possible to predictable rates of deterioration.

2. *An emergency system* where demand for work is essentially unpredictable and action results from the receipt of a complaint or a demand for service. The method of planning maintenance is to define clearly what tasks are to be carried out and then to programme the work in a logical pattern. If this is done properly, failures will be detected at an early stage, enabling simple corrective action to be taken and the level of productivity of the work force will be increased by the logical concentra-

tion of tasks to avoid unnecessary travelling, re-erection of scaffolding, and waiting for access, other trades or spare parts. Unplanned maintenance may seem to give its attention essentially to those items most in need of it, but it conceals the wasteful elements of the frequent reallocation and movement of resources from place to place to deal with tasks of a work content which could have been greatly reduced by forethought and a planned preventive maintenance policy.

The characteristics of maintenance make it impracticable to compile any long-term programme of operations which is either accurate or comprehensive and this high degree of uncertainty must be recognised in any programme which must have built-in flexibility to cope with the inevitable variations which will occur. The main factors which reduce the level of accuracy in planning maintenance work are:

1. Short duration tasks by small groups of operatives often in diverse places.

2. Complexity of co-ordinating work on several sites by a number of different trades, which are interdependent and not necessarily sequential or balanced in their timing or work content. Thus delays may not only affect the site of their origin but have marked repercussions at other work places.

3. The extent of the work and hence its resource requirements is frequently not known until a preliminary investigation has been carried out. It may then prove to involve a sequence of operations leading to requests for additional work to be brought forward in the programme.

4. Programmes usually embrace not only groups of buildings but also several locations and this may mean that efficient labour utilisation can only be achieved at the expense of moving resources between sites more frequently than is desirable.

5. The unpredictability of requests requiring urgent attention and the need for immediate resource allocation. Whereas some failures can be predicted to accompany certain conditions (burst pipes and frost, or roof damage and gales), the very occurrence of these conditions cannot be easily predicted.

6. Delays, additional to those noted in the chapter on contract planning, may be caused by:

    (a) the need to withdraw men to deal with emergencies or to carry out critical tasks elsewhere which have slipped out of phase,

    (b) the non-availability of materials or spare parts required to match or replace those existing.

In view of these uncertainties, programmes must be drawn up at different levels and covering several time spans. These levels are usually defined as:

Long-term   —   quinquennial or longer
Medium-term   —   annual or biannual
Short-term   —   monthly or less.

Long-term programmes

These identify the major items of work to be undertaken during the next five years or longer, and are formulated upon past records of major repairs or redecorations and reports of the current state of the various elements.

These programmes usually contain two main features, the painting and decorating cycles and major repairs. The former has the characteristics of predictability, though the length of cycle will vary in accordance with the location and use from, say, 3 months in an extreme case to seven years. Major repairs, involving replacement or renewal of elements or components, usually result from a desire to eliminate areas of high maintenance cost or trouble which have built up over the years as the need for minor repairs becomes more frequent. The rate of deterioration may be accelerated by failure to take early remedial action, thus changing minor into major repairs. Warning signals are usually present for the experienced eye to detect and thus such work can be fitted appropriately into the programme of work.

Medium-term programmes

It is usual to programme work on an annual basis to relate to the financial budget predictions. Such a programme would be made up of:

1. Items from the long-term plan.
2. Definable items revealed from inspections as requiring attention during the coming year.
3. Undefinable items shown from inspections as likely to be requiring attention.
4. Day-to-day maintenance items based on the experience gained over previous years.

The items will then be costed, assessed for their resource requirements and allocated to the direct labour force or to outside contractors, time being allowed in the programme for tendering procedures in the letting of the work.

In deciding in which order the work will be carried out consideration will be given to the pattern of user needs and steps taken to avoid interference wherever possible, or preparatory arrangements programmed for to provide alternative temporary facilities. The aim should be to ensure a continuous flow of work for the direct labour force as possible without undue recourse to overtime working.

The methods of planning the work by Critical Path or Bar Chart techniques as illustrated in Chapter 5 should be used for the annual overall programme, together with labour resource schedules to highlight peaks of demand in each trade, indicating how these may be adjusted by varying the amount of work let on contract.

Short-term programmes

These are used to provide the flexibility in the programme and to set out in detail the sequence of work on each job, the duration of each operation and the labour, plant and material resources needed to carry it out in the case of direct labour. It will depend on the size of this force as to the extent of the detailed programming required, and the critical chart will be that which shows the planned deployment of the whole labour force on a weekly, or even daily basis.

The problem arises as the mix of trades varies and demands grow out of balance, leading to queuing and delays within the sequence as one trade is overburdened and others underemployed. The degree of flexibility which it is possible to introduce is also limited by the variations in the size of jobs, which to a large extent govern priorities as much as the urgency of a repair job itself.

Thus it will be realised that the programmes act as much as financial predictions for the budgeting purposes, as indications of the sequence of work for a year ahead. With the use of Critical Path Analysis it should be quite possible to examine the consequences of changing the order in which jobs are scheduled, such retiming possibly proving the key to releasing scarce resources for other work.

The execution of the work

Minor works of repair, maintenance, alteration or improvement may be carried out by either contracting firms or by the building owner's directly employed labour force. In both of these cases the problems of organisation and control are basically similar, though the details of how they may be dealt with vary according to the standpoint of the contracting organisation and the nature of the work which they are undertaking.

The major problems likely to be encountered are:

1. Obtaining sufficient continuity of work to maintain a labour force as an economic unit.

2. Attracting and keeping experienced tradesmen.

3. Organising and controlling the resources necessary to carry out small items of work often in geographically diverse locations.

Due to their heavy capital investment and high overhead costs, many large contractors find it difficult to make an adequate return to justify undertaking this type of work. Furthermore, their organisational structure cannot easily be geared to provide the flexibility which is necessary to meet the varying demands of work.

However, some large contractors have small works divisions which not only carry out small new works and alterations but also redecoration and maintenance. Many of these firms may have a long-standing association with large industrial or commercial clients and may undertake all their minor works programme rather as though they were themselves part of the client's own organisation even to the extent of providing a twenty-four hour emergency service as they virtually have a permanent labour force employed on the premises.

The choice of a method

The main decision which the maintenance manager may be required to take is whether to employ a direct labour force or to contract the work outside. The choice will probably depend upon a balance between cost and convenience to the users, bearing in mind the availability of directly employed labour and the programme of work to which he is committed.

In making his decision the manager will be influenced by several factors, not least whether the organisation is likely to generate a workload of sufficient size to justify maintaining a permanent direct labour force. The decision must be based upon an economic evaluation of the alternatives and an assessment of the relative productivity likely to be achieved as it will be necessary to keep a directly employed labour force occupied at a productive level for the working year if its cost is to be comparable to that of an outside contractor. The cost of such a labour force will be determined not only by the wage bill of the employees but also by the overheads necessary to keep the organisation in existence and operating effectively. This, added to plant and material costs when allocated to the maintenance work that is carried out, will give the true picture of cost, which must be compared, from time to time, with quotations for similar work from outside contractors. This in turn can be converted into "productivity data".

There are, however, other factors which will influence the choice between the two alternatives and in many cases the direct cost may well be misleading as a main determinant when the hidden costs of, say, interruption to industrial production or hazard to life or structure are taken into consideration. Thus it is necessary to compare the alternatives in the light of these other factors.

Directly employed labour

It has been suggested above that whereas the use of direct labour may in some cases show a cost benefit to the organisation, in others it may well result in an economic penalty which will outweigh the direct cost advantage. The factors which may prove influencial are listed below.

Response time

A permanent labour force may be considered as an insurance policy to guard against failures in a building or its services which may have cost consequences other than those of the resultant repair. Loss of a facility is often difficult to quantify in financial terms, though in cases of industrial production or retail turnover it is comparatively easy. The effects of disruption to clerical work, educational establishments or health services usually cannot be predicted so readily and may need to then be judged upon past experience. Where failure in the supply of a service may result in a hazard to life, as in a hospital or chemical plant, the criterion is one of safety and cost considerations are of minor importance and can be judged only against the likelihood of such an eventuality occurring.

Security

There are situations where security is an important consideration and it may be desirable to restrict access to premises as far as possible. It is generally possible for directly employed labour to be vetted and controlled more readily than men employed by outside contractors, thereby safeguarding to a greater extent commercial and other classified information.

Familiarity

There is some advantage to be derived from a work force which has become familiar with the layout of premises and their services and is therefore more likely to recognise quickly the causes of failure from their continuing association with one employer's estate, than in the case of externally employed contractors who are called in only as troubles occur and will not necessarily be able to send men who have knowledge of the premises. A permanent labour force can also be trained in the maintenance of a particular group of installations and may be sent on specialist courses in the knowledge that the training will have direct practical application.

Such advantages, however, can be overcome by the system of "Term Contractors" or other special arrangements with individual contractors.

Control

One advantage which arises from the direct employment of labour is that the nature of the work and the security of employment which it implies often attract the mature tradesman, with consequently a reduced labour turnover. This can create the opportunity for greater flexibility in the deployment of the labour force and smoothing out of the maintenance programme. If one task is delayed for any reason it is comparatively easy to avoid non-productive periods by switching labour to other work. However, there is always danger that this very flexibility may be seen as an excuse not to plan work rigorously and to rely upon "instant management" techniques to overcome such eventualities. The apparent full employment of staff may hide the additional costs which are always incurred by any description of a well planned programme, particularly when it affords the opportunity to bring items forward, often at the expense of more urgent work.

254

Even the ready availability of labour to deal with emergencies may dissuade the regular inclusion in a programme of measures which are designed to minimise the occurrence of such emergencies.

Nevertheless, the facility for control, which is a part of this method, may make it better suited to routine tasks and preventive maintenance, whilst ensuring a rapid response to night calls. Furthermore, familiarity with the work patterns of the users of a building can lead to less disruption and hence less potential financial loss during the actual execution of the work itself.

It has already been stressed that a master programme of study showing the deployment, actual and planned, of the whole labour force is essential if full utilisation is to be obtained and some degree of flexibility maintained between jobs.

New work

Where it forms a substantial part of the employing organisation's building programme, minor new work, either in the form of extension or adaption and alteration, is likely to be carried out by directly employed labour and this may be seen as a good opportunity for adding to the flexibility of the work force. Unfortunately the maintenance manager may find that it leads to his losing any direct control of the labour resources and maintenance may be looked upon as a useful buffer in periods of underemployment of building labour, whilst in periods of peak activity he may find no labour at all available to him.

In the first case he may be required to find employment even when he has no requirement, and in the second he will find it impossible to proceed with his programme of maintenance work. He may also find it increasingly difficult to ensure that the particular type of craftsman is available when required. It is essential to create a balanced labour force to meet the requirements of both activities in terms of number and the trade skills represented. This also implies a close relationship between the programme of maintenance and new work.

However, with good planning and co-ordination of resources and responsibilities such a combination of activities can prove successful and there are many contracting organisations which have deliberately incorporated an "odd jobbing" or "maintenance and repair" section to provide just this means of balancing their overall work load by introducing a more stable activity to counter the greater fluctuations of the general building market.

In assessing the real cost of a direct labour organisation the overheads must be realistically taken into account and the size of the organisation needed to function effectively. The maintenance department must expect to require a similar organisation to that of a contractor if it intends to carry out work itself. Provision must be made of:

1. Supervising staff experienced in the execution of minor works and repairs.

2. Facilities for the storage of a stock of materials and components in general use with adequate purchasing and stock control.

3. Basic items of plant, tools and equipment, together with covered storage space. It is envisaged that larger items of mechanical plant, if needed only occasionally, will be hired. A stock of scaffolding will be kept sufficient for everyday maintenance requirements, whilst any major needs for a particular task will probably be hired.

4. Transport for both men and materials.

5. A small workshop unit for servicing plant, equipment and vehicles and carrying out basic workshop operations.

6. Administrative support staff for such functions as wages and estimating under the control of the maintenance manager and possibly his deputies, housed in appropriately appointed office accommodation.

The employment of external contractors

The alternative to employing direct labour is to engage external contractors to undertake the work and these may be used for general maintenance work or in a specialist capacity, and should be considered separately as the contractual conditions may vary in each case.

General work

Many large industrial and commercial organisations choose not to commit themselves to a permanent maintenance staff but prefer to rely upon outside contractors, very often on a continuing basis so that there is usually some of the contractor's labour constantly employed around the estate. This method also has attractions for organisations with a scatter of individual buildings over a wide area where associations can be established with local firms and the problems and expense of moving resources from a central depot can be avoided.

There are a number of contractual systems available to the maintenance manager for letting out work, each appropriate to certain conditions. Too often little thought is given to the work until it actually arises, when an order is given to a small local builder usually on a daywork basis. This can prove a very costly way of getting maintenance work carried out and a good manager will not only programme ahead but also have a constant policy regarding the contractual procedures to be used. In general, some form of measurement contract can prove more economic if staff are available to measure the completed work, as it provides an incentive to the contractor to complete the work quickly.

The main types of contract usually available for day-to-day repairs in preventive maintenance are:

Daywork

This may be based either upon a fixed hourly rate for each craftsman or upon a fixed percentage for overheads and profit on the actual rate paid to different categories of craftsman, ie, a variable rate for each man. A careful check of the time occupied on particular items of work must be made if this method is not to prove unduly expensive.

Schedule of rates

This may consist of a comprehensive prepriced schedule of rates with a full specification of the work required with plus rates provided by the contractor and updated at prearranged intervals. Special schedules may be provided to cover a specific range of work which is normally encountered in a particular group of buildings.

The DOE schedule provides an example of a standard prepriced schedule containing some 6000 items of which a contractor will probably select only the predominant trades and base his percentage on the estimated quantities of the major items in these trades.

Thus whilst it is a convenient method of pricing work of conventional scope, it lacks estimating accuracy. However, it establishes standard work descriptions and lays down a pattern of pricing which is sufficiently reliable to form an acceptable basis for valuing work executed. Percentage additions also provide an indication of trends in maintenance prices.

Term contracts

This type of contract is designed to afford some measure of continuity of association between a client and a contractor by providing the opportunity to carry out all work of a certain type or within a price bracket for an agreed period. It is usually based upon a priced schedule or a system of cost reimbursement with a comprehensive specification covering all work likely to be encountered and is usually let on a competitive basis by contractors offering a percentage offer on the completion schedule. They are usually most successful when applied to certain closely defined repetitive small jobs and when the period covered is not too long and increased costs can be reasonably forecast. One problem is that great reliance must be placed upon the quality of the written works instruction or job description given by the maintenance manager's staff as this forms the ultimate basis for payment, and this is why some contractors are hesitant in undertaking work under this form of contract. However, it enables a contractor to gain experience of working at a particular estate and fitting his own plan into the current programme. Emergencies can also be covered comparatively easily as he will have men available either already on the estate or probably aware of its layout. Daywork term contracts may be more suitable for organisations with a scatter of individual buildings as completion of these is subject in percentage form requiring figures for labour and materials and transport costs. These additions are the tenderer's requirement to gauge profit, expenses and overheads, whilst labour and materials are paid for on the basis of time-sheets. One problem for the client is that any slackness on the part of the contractor on site is unlikely to incur him in any loss as it will be reflected only in additional hours worked; thus a close control must be kept on the execution of the work.

Special work

In certain cases where work of a specialist nature is involved, a pre-priced schedule of rates may be inappropriate, though term contract arrangements are desirable. In these cases it is usually possible to draft contract conditions which relate specifically to the work required, the likely frequency of attendance and such normal conditions as may be relevant. Tenders can then be obtained in the form of a price per visit or unit of work. This is appropriate for such work as window cleaning, boiler and lift maintenance, inspection of pressure vessels, masts and towers, and the maintenance of grounds and gardens. It will be realised that several of these items carry a statutory obligation for periodical testing which can often be obtained through the services of the company issuing the equipment; the inspectors will be recognised as "competent persons" in terms of the statutory obligation.

Summary

The management of maintenance work can thus be seen to have specific problems originating in the nature of the work, though the techniques of costing and production control which have already been outlined for general construction are applicable once these problems have been identified. Those characteristics which have the most important influences are highlighted in the following points:

1. Maintenance work relates to existing property which is normally occupied whilst the work is executed.

257

2. Maintenance work therefore usually involves a building's occupants, who are liable to be deprived of some amenity, occasioning inconvenience and calling for good personal relations.

3. Finance is usually minimal as first costs do not cover maintenance which is taken out of current budgets. The less the money, the more the troubles that will accumulate.

4. Work is rarely of a major or continuing nature — there are few Forth Bridges to be painted! — but there is increasing awareness of the need for long-term programmes. Maintenance manuals for new buildings encourage this.

5. Emergencies are always short-term and expensive in resources.

6. Incentives are harder to apply to this work but may be possible through the application of method study.

7. Work is of short duration carried out by small groups of labour often at a distance from their base.

8. The full problem is not often known until work has started, so that contingency allowances need to be large to allow for the unexpected. The time spent in fault diagnosis alone is often larger than is realised.

9. Maintenance is not often considered at the design stage; or, if it is, design faults can undermine the forethought.

10. Term contracts or directly employed labour may enable rhythm and planning to be introduced with a balanced labour force.

11. A range of standards will be needed and the quality of work must meet the appropriate level, in some cases for statutory reasons.

12. Feedback of experience is essential and must be in terms of building types, defects, quality standards, etc. Information should be recorded in standardised comprehensive form to ensure its classification and retrieval.

13. Records are very necessary to provide the basis for inspection routines and the preparation of programmes related to building and component life cycles.

14. Maintenance work calls for a wide variety of skills which are becoming increasingly scarce.

15. The role of maintenance manager is becoming increasingly demanding and requires a balance of technological expertise and practical experience, together with a training in analysis and an understanding of its role in the overall performance of an organisation.

BIBLIOGRAPHY

1 LEE, R. Building maintenance management. Crosby, Lockwood, Staples, 1976.

2 ROBERTSON, J A. The planned maintenance of buildings and structures. The Institution of Civil Engineers, Proceedings paper 7184S, 1969.

3 Conference papers of building maintenance conferences.

4 THE INSTITUTE OF BUILDING. Maintenance management: a guide to good practice. The Institute of Building.

Appendix

JD1

Job title: Head of Contracts Department

Visits the site at least once a month with the Contract Manager. He provides a link between the project and the senior executives; since he is a member of the Board of Directors he also provides a link between the Project and the Board.

As with all projects he expects a continuous flow of control information relating to finance and production. In this particular case he maintains contact with the client's senior representative.

JD2

Job title: Contracts Manager ✓

1. General job description

Daily
    (a) Answering correspondence and any queries and distributing information.
    (b) Contacting site and solving any problems which arise.

Weekly
    (a) General planning of new contracts.
    (b) Updating programmes and assessing work carried out.

Monthly (or as may be necessary)
    (a) Meetings with architect and sub-contractors.
    (b) Reviewing correspondence of each contract.

General
    (a) Organising new contracts and administering contractual obligations.
    (b) Organising the progress of contracts.
    (c) Obtaining maintenance defects lists on completed jobs and ensuring listed work is carried out.

2. Qualifications and training
MIOB preferred with a good basic knowledge of the technical and contractual aspects of the industry.

3. Experience
    (a) Not less than 4-5 years experience preferred in site managerial or surveying position.
    (b) Knowledge and experience of Head Office organisation with particular emphasis on surveying and planning.

4. Special skills
    (a) Leadership qualities.
    (b) Ability to assess possible problems; to work under pressure and to work on own initiative.
    (c) Must be both physically and mentally fit.
    (d) Must be able to get on well with people.

5. Responsibilities
   (i) (a) Responsible to head of contracts department.
       (b) Responsibility to and authority over site management.
   (ii) Responsible for:
       (a) To the best of his ability and limits of authority to produce the budgeted profit for the contracts under his control.
       (b) Maintenance of customer goodwill.
       (c) Keeping proper records relative to all production.
       (d) Studying and recommending use of new managerial and technical techniques.
       (e) Training of supervisory site management.
       (f) Co-ordinating use of labour on various sites.
       (g) Co-ordinating the use of plant on various sites.
       (h) Ensuring the completion of contracts on time (allowing for extensions, etc.)
       (j) Ensuring that income exceeds expenditure.
       (k) Expenditures, materials; labour; plant, entertaining clients, etc.

6. Contact with others
   (a) Frequent contact with clients, customers, architects, suppliers, sub-contractors etc.
   (b) Frequent contact and co-operation with other members of staff, e.g. surveyors, buyers, plant manager, etc.

7. Adverse working conditions
   (a) Safety hazards on site.
   (b) Adverse weather conditions on site.
   (c) Necessary to work late.

8. Errors that could have a detrimental effect on the company's business.
   (a) Wrong decisions relating to programming.
   (b) Wrong decisions relating to safety precautions.
   (c) Setting-out errors.
   (d) Failure to execute contractual responsibilities.

9. Limitation of authority
   (a) He shall have the authority to spend up to £x per week for necessary items such as plant materials, etc., providing the purpose of the expenditure is to avoid a production delay. For expenditure over this amount he must first obtain the approval of the Head of Contracts Department.
   (b) He shall have the authority to engage or dismiss all site employees (except site management and craft foremen) for the ordinary execution of work; but in the event of any proposed dismissal related to a breakdown in industrial relations he must first report to the Head of Contracts Department and the Industrial Relations Officer.

JD3   **Job title: Site Manager** √

1. General job description
   (a) Immediately on appointment familiarise fully with all details of the tender documents.
   (b) Visit site, agree staffing and accommodation schedules with the Contracts Manager.
   (c) Senior company representative (unless the Contract's Manager present) at all meetings with the client and architect. At the first meeting would present Master Programme for Architect's approval and also 'Information Required Schedule' (Figure 5). At this meeting it would not be unusual for the Architect to issue the Contract Drawings.
   (d) Arrange pre-contract meeting at the Contractor's head office to discuss site policy and procedures with senior staff.
   (e) Arrange pre-contract meeting to discuss methods statements with Site Managers, General Foremen, Planner, Safety Officer, Materials Controller.
   (f) Arrange whatever meetings seem necessary during the construction period but the general case would be to arrange a company meeting of senior site staff before the Architect's monthly site meeting which he would attend.

Daily
  (a) Look at programme and take action if necessary.
  (b) Examine correspondence in and irregular spot checks on correspondence out.
  (c) Check progress of issue of information and drawings.
  (d) Check new and revised drawings as issued.
  (e) Walk round the site.

Weekly
  (a) Meet with site managers to discuss progress.
  (b) Meet shop stewards or job steward to discuss matters arising.
  (c) Meet Office Manager to check timekeeping and attendance.
  (d) Check the quality of labour and check the quality of work done.
  (e) Check on plant utilisation.

Monthly
  (a) Meet the Architect.
  (b) Chair monthly site meeting.
  (c) Check financial progress of project with Senior Quantity Surveyor.
  (d) Check Production progress with Site Managers.

General
  (a) Senior representative of the company on site.
  (b) Ensure that work is being done to the required quality standards.
  (c) General supervision of safety, security, and administration.
  (d) Investigate any complaints which come to him through proper channels.

2. Qualifications and training
  (a) MIOB or degree in Building studies.
  (b) Chartered engineer status if the engineering aspects of the projetc outweigh the general building aspects.

3. Experience
  (a) Ideally will have worked successfully for more than one firm as a site manager. The size and complexity of work undertaken to date will be appropriate to the position now occupied.
  (b) Knowledge and experience over a wide range of activity including planning, general foremanship, quantity surveying.

4. Special skills.
  (a) Leadership qualities including optimistic confidence in his own ability, highly motivated to get on with the work; he must be a team builder able to draw the best out of everyone.
  (b) He must recognise and practice the management processes (as identified by Fayol).
  (c) Must be both physically and mentally fit.
  (d) Must be able to get on with people.

5. Responsibilities
  (a) Charged with the responsibility of bringing the project to a satisfactory conclusion from Client and Contractor viewpoint.
  (b) Must maintain fair, ethical standards of employment for all engaged on the project.
  (c) Must accept responsibility for everything which occurs on site.

JD4   **Job title: Senior surveyor**

1. General job description
  (a) Monthly cash forecasts.
  (b) Assessment of turnover and overheads and profit recovery.
  (c) Attend monthly finance meetings.
  (d) Attend monthly staff meetings.
  (e) Annual work-in progress.

2. Tasks performed
Daily
  (a) Ensure that correspondence and records in connection with contract receipts and payments are kept in an up to date condition.

(b) Keeping records of all delays caused by the late receipt of design information and other relevant data, the late attendance of sub-contractors, and other incidents outside the company's control.

(c) Liaise with other departments in the provision of operational controls in respect of the progress of contracts and other works.

(d) Advising management of any difficulties arising in connection with the agreement of claims, interim and final valuations, or of conditions imposed after work has commenced.

(e) To safeguard all monies, property, documents, records and other information belonging to the Company, its clients or personnel.

Weekly and Fortnightly

Measurement of work done by labour only and appropriate domestic sub-contractors, checking and authenticating before certifying payment.

Monthly

(a) Ensure that detailed site measurements of works in progress (including sub-structures, services, finishings and external works) are carried out.

(b) Detailed cost/valuation comparisons should be prepared within deadline time of agreeing each valuation and statements presented to management as required.

(c) Interim and final payments and accounts payable to own and nominated sub-contractors and nominated suppliers should be checked and authenticated.

(d) Interim applications for payment on account of work done should be made at the correct time and agreed with clients or their agents.

Occasionally.

(a) Attending site progress meetings where necessary.

(b) Ensuring that all contract documents, conditions, drawings, design, information, bills of quantities and other information are carefully examined; taking notes of any onerous terms, risks, etc., for action where necessary.

(c) Preparing reports on examinations of contract documents for discussion at pre-contract meetings and during mobilisation stages of contract work.

(d) Ensuring that final accounts, fully complete as to variations, additions and extras are submitted and agreed.

3. Qualifications and training

MIOB and or AIQS preferred with a good knowledge of all forms of building contracts and SMM.

4. Experience

Approximately 10 years experience and with a good general knowledge of all aspects of the building industry.

5. Special skills

(a) Must hold a current driving licence.

(b) Must be both physically and mentally fit.

(c) Must be a good negotiator.

(d) Must be able to communicate both verbally and by writing.

6. Responsibilities

(i) (a) Responsible to Head of Contracts Department

(b) Responsible for two surveyors and one assistant surveyor.

(ii) Responsible for:

(a) Assets: Company car.

(b) Expenditure: Payment of sub-contractors and agreement of final account.

(c) Income: Agreement of valuation and final account with QS.

7. Contact with others

(a) Frequent contact with clients, quantity surveyor, architects, suppliers, sub-contractors, etc.

(b) Frequent contact with other members of staff, eg surveyors, contract managers, buyer.

8. Adverse working conditions

(a) Safety hazards on site.

(b) Adverse weather conditions on site.

(c) Necessary to work outside regular company hours.

9. Errors that could have a detrimental effect on the company's business

Omissions and mistakes when agreeing valuations, claims, extension of contract, etc. with Q.S.

Job title: Surveyor

1. General job description
Daily
    (a) Answering correspondence
    (b) Dealing with day-to-day problems and routine surveying duties.

Weekly
    (a) Visit sites to familiarise with progress and anticipate problems which could arise.
    (b) Measurement of work done by labour only sub-contractors and certify payment.
    (c) Keep daywork sheets and weekly records of labour up to date.
    (d) Measure direct labourers' work and apply bonus targets, evaluate and certify payment.

Monthly
    (a) Agreement of valuation with QS including attention to day works and variations.
    (b) Payment of sub-contractors.
    (c) Forecast of monthly valuations to help anticipate cash flow.

General
    (a) Measurement and agreement of final accounts.
    (b) Obtaining quotations for sub-contract work.
    (c) Placing orders with sub-contractors.

2. Qualifications and training
    HNC/HND with knowledge of forms of building contracts and SMM.
3. Experience
    2 years experience preferably involved with general surveying duties.

4. Special skills
    (a) Ability to write letters.
    (b) Ability to work with figures.
    (c) Must be physically fit.
    (d) Must hold a current driving licence.
    (e) Must be capable of working with the minimum supervision.

5 Responsibilities
    (i) (a) Responsible to senior surveyor.
         (b) Responsible for assistant surveyor.
    (ii)   Responsible for:
         (a) Assets: Company car, surveyor's measuring equipment.
         (b) Expenditure: Payment of sub-contractors.
         (c) Income: Agreement of valuations with QS.

6. Contact with others
    (a) Frequent contact with sub-contractors.
    (b) Frequent contact with other members of staff.
    (c) Frequent contact with Client's Quantity Surveyor.
    (d) Occasional contact with Architect.

7. Adverse working conditions
    (a) Safety hazards on site.
    (b) Adverse weather conditions on site.

8. Errors that could have a detrimental effect on company's business.
    (a) Over-payment of sub-contractors.
    (b) Omissions and mistakes when agreeing valuations with QS.

Job title: Assistant surveyor

1. General job description
    (a) Filing letters and drawings.
    (b) Pricing dayworks.
    (c) Preparing payments of labour only sub-contractors.
    (d) Preparing increased costs for materials and labour.
2. Qualifications and training
    School-leaver preferably with 'O' levels and with intention of continuing education by studying for Higher Technicians Certificate.

3. Experience
   None

4. Special skills, etc.
   (a) Ability to use surveyor's rod and tape.
   (b) Must be numerate.

5. Responsibilities
   Responsible to surveyor.

JD7

Job title: General Foreman

1. General job description
Daily
   (a) Organising and supervising work including sub-contractors.
   (b) Ensuring that only safe methods of work are being used and withdrawing equipment which is unsafe.
   (c) Checking that all work is being done to the required quality standards.
   (d) Arranging the erection of scaffolding, formwork, and obtaining the necessary equipment.
   (e) Communicate instructions received from the site manager to the sub-contractors foremen and company craft foremen.
   (f) Maintain hoardings, gates, and pavements in a safe, clean condition.
   (g) Maintain a site diary.
   (h) Maintain adequate records for all work done under his control.
   (j) Set out the Works.
   (k) Arranging and supervising taking off operational quantities.
   (l) Scheduling and calling forward materials and plant.
   (m) Making recommendations to the Site Manager concerning recruitment selection and engagement of labour, instigating normal termination or dismissal procedures after consultation wth the Site Manager.

Monthly
   (a) Planning stage programmes in conjunction with the Planner.
   (b) Plan site lay-outs, including materials and stores location, check stock levels of 'stored' materials.
   (c) Attend site meetings as called by the Site Manager.
   (d) Operating and controlling the bonus scheme in collaboration with the Production Surveyor.

2. Qualifications and training
   (a) HND or HND with at least two years experience or alternatively a person with good trade background who has studied adequately beyond the limits of his own craft, or a person with Construction Technicians training who has completed the Part II studies and has not less than 3 or 4 years site experience.
   (b) Training should include setting out and levelling, and management, quality control of building work.

3. Experience
   Should have worked on a number of different contracts preferably with different site managers.

4. Special skills, etc.
   (a) Able to motivate and supervise men at work.
   (b) Communicate clearly by verbal means to craft foremen.
   (c) Ability to think clearly with particular regard to methods of work and to forecast needs before they arise.
   (d) Emotionally stable
   (e) Read and understand drawings.

5. Contact with others.
   (a) Daily contact with the Site Manager and the craft foremen and sub-contractors foremen.
   (b) Meeting as required with the Safety Officer, Security Officer, Production Surveyor, Quantity Surveyor, and Project Manager.

JD8   **Job title: Senior production surveyor.**

1. General job description
   (a) Establishing pre-determined standards of performance for work operations, these having a relationship with the estimator's rates, and translating these standards into practical bonus target rates for operatives.
   (b) Physical measurement on site of work done both for cost control and bonus payments due.
   (c) The comparison of the actual performance both in detail and total with the pre-determined standards.
   (d) Collaboration with Site Managers to determine the reasons for differences between planned and actual performance levels of productivity.
   (e) Critically reporting to the Project Manager on work activity including:
       (i)   the use of plant, gang balance, techniques.
       (ii) the true cost of 'spot' cost items.
       (iii) activity sampling carried out.
   (f) Determining the true cost of employment of labour-only and labour and materials sub-contractors and comparing this with company employed workers.
   (g) In conjunction with the Quantity Surveyor recording Variation Orders, Dayworks, and Site Works Orders.
   (h) Liaising with the Materials Controller to measure, control, and where possible reduce wastage of materials.
   (j) Make comparative studies of the types of plant that can be used in operations.

   Daily
   Collect daily labour and plant allocation sheets.

   Weekly
   (a) Prepare weekly cost sheets.
   (b) Prepare weekly bonus sheets.
   (c) Compare actual costs with planned costs and assess which of the variances are significant.
   (d) Discuss significant variances with Site Managers and co-operate in seeking solutions to problems.
   (e) Check with Materials Controller on material reconciliation.

2. Qualifications and training
   Construction Technician Part II qualification with special training in Work Study and bonus scheme operation.

3. Experience
   One year assisting in Buying and two years as assistant to General Foreman.

4. Special skills
   (a) Ability to work with figures.
   (b) Must be able to work and plan his own work without reference to higher authority.
   (c) Must be logical, critical, unbiased.
   (d) Must be able to use a computer.

5. Responsibilities
   (a) Immediate superior is the Project Manager.
   (b) Immediate sub-ordinate is the Assistant Production Surveyor.

6. Contact with others
   Weekly contact with the Site Manager and frequent contact with the Materials Controller and General Foreman.

JD9   **Job title: Senior engineer**

1. General job description
   (a) Receive all drawings concerning structure and maintain an up to date set with superseded copies marked and stored separately.
   (b) Check all drawings when received and notify the Contracts Manager immediately of any discrepancies.
   (c) Set out in accordance with the drawings all work, check that the structure is being erected to the setting out and in the event of any discrepancy notify the Project Manager immediately.

265

(d) Maintain field, level, and calculation books in such a manner that they can be immediately interpreted by any other Engineer.

(e) Maintain instruments in a clean and accurate condition, use them carefully, and keep them safely.

(f) Operate a system of checking all work before concreting commences.

(g) Prepare test cubes as necessary and arrange testing.

(h) Supply levels, dimensions and sizes as may be reasonably required by sub-contractors.

(j) Design and supervise any authorised temporary works necessary to the Project.

(k) Liaise with the Architect and Consultant Engineer.

Daily

(a) Maintain a comprehensive daily diary.

(b) Maintain daily records of progress by marking up the appropriate drawings.

Monthly

Attend site meetings if required.

2. Qualifications and training

HNC/HND background with structural bias. Keep up to date with relevant Codes of Practice, Formwork design, design of structure. MIOB preferred.

3. Experience

Four years construction site experience.

4. Special skills

(a) Ability to teach trainess how to carry out the duties of a site engineer.

(b) Work accurately without supervision.

(c) Plan work to be in advance of the needs of the construction team.

5. Responsibilities

(a) Responsible for the accurate setting out of the building.

(b) Responsible for reinforcing rod being to the right size and positioned correctly.

(c) Responsible for the quality of the concrete.

JD10   **Job title: planner**

1. General job description

(a) Liaise with sub-contractors concerning their period of attendance on site before preparing any programme.

(b) Prepare a realistic overall programme for the Project.

(c) Prepare overall programmes for the different sections of the Project.

(d) Assist General Foremen in the preparation of stage programmes.

(e) Check that drawings, schedules, etc. are being received in good time from the design team and specialist sub-contractors and suppliers.

(f) Record the receipt of drawings, etc. and arrange their distribution.

(g) Mark up the overall programme weekly, send weekly progress report to Contract Manager.

(h) Send feed back information to head office Planning Section from which they may derive constants for use on future projects.

2. Qualifications and training

HNC/HND General site experience, work study/costing.

3. Experience

2 years' experience as mentioned above.

4. Special skills

Must have a critical and at the same time imaginative approach to work.

5. Responsibility

Subordinate to the Site Managers and responsible for preparing their long term and stage programmes. In some cases he may help the General Foremen to prepare the weekly programmes.

Job title: Services co-ordinator

1. General job description
   (a) On receipt of any building services drawing check the
       other services relative to location, timing of its ins†
       structural members.
   (b) Wherever possible pre-determine the position of all ⊦
       such as floor members, arrange the orderly grouping
       these floor units and arrange if possible for pre-formed
   (c) Check the size, weight, delivery arrangements, of
       equipment, plant, etc. and arrange receipt, access, hoisting into posiuun, ⌐⁻⁻
       ion, etc.
   (d) Ascertain sub-contractors builder's work requirements in advance of construction
       programme.
   (e) Make certain that any services task which could affect the critical path of the pro-
       gramme is fully discussed and analysed with the relevant sub-contractor and that
       they are fully aware of the need to control their activity and keep to programme.
   (f) Notify the supervising authority of any deviation in design and obtain Variation
       Order where necessary.
   (g) Control any Daywork or additional work items carried out by sub-contractors
       which could be charged against the main contractor. Similarly notify the sub-
       contractors where contra-charges may be placed against them.
   (h) Check reported progress figures for accuracy.
   (j) Attend all site meetings to give 'building services reports'.
   (k) Check that all valves and regulating units are correctly labelled and that this
       information is recorded for handing to client.
   (l) Be available when any installation is being commissioned.
   (m) Organsie meetings as required for services trades.

2. Qualifications and training
   HNC/HND

3. Experience
   3 or 4 years site experience with specialist services firms and some design experience
   desirable.

4. Contact with others
   Regular consultation with Site Managers' sub-contractors, the Planner, to ensure
   maximum co-ordination.

Job title: Materials controller

1. General job description
   It is assumed that a Materials Controller will be appointed on large sites only. Large
   sites are frequently divided into sections, with a site manager appointed in charge of
   each section. The Materials Co-ordinator plays a major role in the ordering, supply and
   distribution of all materials.
   (a) Taking off materials in conjunction with Site Managers. Investigating any queries
       which may arise at this stage regarding security of supply, packaging, delivery,
       handling, storage, and usage.
   (b) Ordering materials through the Contractor's Buying Department for all site
       requirements.
   (c) Monitoring supplies in and 'chasing' materials when production delays on site are
       likely to occur because of non-arrival.
   (d) Discussing administrative, delivery, or fixing queries, with suppliers of specialist
       equipment, fitments, fabricated items, e.g. structural steelwork, escalators, lifts,
       joinery, sanitary ware, partition panels, and liaising in these matters with the
       Architect and other Consultants.
   (e) Controlling material waste by improving site handling techniques, planning access
       routes, organising storage, better plant selection, careful programming of
       deliveries and distribution to the sections more careful measurement of materials
       at the user point.

2. Qualifications and training
   (a) MIOB Institute of Purchasing Officers.
   (b) 2 years' training experience in Buying, Plant, and Site Management.

3. Experience
Two years' experience as a Buyer.

4. Special skills, etc.
   (a) Ability to plan ahead and forecast problem areas.
   (b) Must be a good negotiator.
   (c) Essential that contact is maintained with market trends, innovations in plant and handling equipment.

JD13   **Job title: Security Officer**

1. General job description
   (a) Main efforts of the Security Officer are directed to the prevention of theft. Four main groups for consideration are:-
      (i)   persons not in the employ of the company and not engaged on the site.
      (ii) persons not in the employ of the company but having legitimate reasons for being on the site.
      (iii) persons in categories (i) and/or (ii) acting in collusion with company employees.
      (iv) company employees.
   (b) Minimise losses due to carelessness.
   (c) Guard against damage by vandals.
   (d) Give special consideration to the site according to its particular vulnerability possibly because of site locality, or the site lay-out, type of construction, or mode of employment of labour.
   (e) After the tender stage survey of the site and study of the site lay-out, establish procedures and routines for:-
      (i)   receipt and checking of incoming vehicles.
      (ii) checking of outgoing vehicles.
      (iii) limiting authority to sign for goods incoming.
      (iv) limit authority to sign internal 'Requisition Notes' authorising goods to be drawn from stores.
      (v)   variable procedures for the receipt on site of wages and the payment of same.
      (vi) checking procedures by foremen.
   (f) Make special provision for the difficult periods like Christmas holidays.
   (g) Make adequate arrangements for the security of workmen's tools as outlined in National Working Rule 3F.
   (h) Where circumstances seem to justify make arrangements for weights and measures checks as appropriate.
   (j) Arrange proper lock-up for fuels, in particular petroleum spirit and set up a tight control system for its issue and justifiable usage.
   (k) 'Educate' operators in the protection of their plant from vandalism.
   (l) Be constantly on guard for fraudulent practices at the 'clocking-in' point and stolen goods hidden under the spoil on 'muck-away' trucks.
   (m) Study the economics of security, ie measure losses against the cost of stopping them.

2. Qualifications and training
   Training in the police would be most suitable.

3. Experience
   On various sites.

4. Special skills
   (a) Capable of training site supervisors, store keepers, material checkers on how to guard against theft and fraud.
   (b) Vigilant.

JD14   **Job title: Safety Officer**

1. General job description
   Regulation 5 of the Construction (General Provisions) Regulations require every contractor who employs more than 20 men to appoint in writing one or more Safety Supervisors. In a large company the Safety Officer has a large involvement in selection/recruitment and training of Safety Supervisors.

268

(a) Advise the Site Manager on the safey requirements of the contract with regard to
    (i)   Factories Acts.
    (ii) Construction Regulations, 160 onwards.
    (iii) Health and Safety at Work etc. Act 1974.
    (iv) Codes of Practice relevant to safety matters.
    (v)   Current thinking of H M Factory Inspectorate.
    (vi) Requirements of the Fire Authorities.
    (vii) Recommendations of the Institute of Industrial Safety Officers.
    (viii)Recommendations of the British Safety Council.
(b) Supervise observance of these legal requirements and promote safe conduct of work.
    (i)   prevent injury to all personnel on site.
    (ii) seek to improve existing methods of working.
    (iii) arrange the provision of protective clothing and equipment, supervise its proper use.
    (iv) advise and train where necessary in the use of new and hired-in plant.
    (v)   check that insurances have been taken to cover all contingencies - particularly such operations as erecting, testing, and dismantling tower cranes (ie not covered by the general use policy).
(c) Arrange proper fire precautions, particularly danger of fuel storage, canteen kitchens, plant parks, certain materials, i.e. roofing felts, polythene sheeting.
(d) Advise management of changes in safety legislation.
(e) Examine unusual construction from the safety aspect.
(f) Examine proposed new methods of working from the safety aspect.
(g) Ensure that proper provision has been made for first aid boxes, ambulance room etc. Liaise with the first aid nurse in these matters.
(h) Advise on the protection of the public on the site boundary and about the site access.
(j) Determine the cause of any accident or dangerous occurrence and recommend means of preventing recurrence.
(k) Supervise the recording and analysis of information on injuries, damage and production loss, assess accident trends and review overall safety performances.
(l) Assist with training for all levels of employee, and suggest posters, slides, films and film strips to promote awareness of injury prevention and damage control.
(m) Carry out site surveys, in association with Site Managers, to see that only safe methods of working are in operation; that all Regulations are being observed; eg that staturtory notices have been posted; that mess rooms, washing facilities and other welfare amentities have been provided and properly maintained.
(n) Take part in site management/operative discussions on injury, damage and wastage control.
(o) Keep up to date with recommended codes of practice and new safety literature; circulate information applicable to each level of employee.
(p) Foster an understanding that injury prevention and damage control are an integral part of business and operational efficiency.

2. Qualifications and training.
   Craft trained from one of the construction trades, Bricklayer, Carpenter.
   Safety Courses organised under auspices of NFBTE, CITB, RoSPA, etc.

3. Experience
   It is essential that he should have had site experience.

4. Special skills, etc.
   Must be able to work at heights, must be enthusiastic about the job and adopt a positive approach to his work.

5. Responsibilities
   (a) Answerable to the Project Manager.
   (b) It is essential that he is outspoken irrespective of seniority aspects if he thinks there is unsafe practices being used.

JD15   **Job title: Office Manager**

1. General job description
   (a) Arrange site administrative procedures to accord with head office procedures both in the information provided and in the correct timing of the provision.

(b) Distribute incoming mail, collect in and maintain record of outgoing mail.
(c) Order all stationery, store, and distribute as necessary.
(d) Set up and maintain site filing system.
(e) Supervise timekeeping, compilation of labour returns, payment of wages.
(f) Receive all visitors and arrange whether they should be seen by a member of the management team or not.
(g) 'Set up' site meetings.
(h) Arrange ordering of canteen foods and give general supervision to the running of the canteen and the cleaning of the office.
(j) Control stores records.

2. Qualification and training
ONC/HND in Administration.

3. Experience
Secretarial and administrative.

Index